MW01517678

GUIDEBOOK FOR WASTE AND SOIL REMEDIATION:
FOR NONHAZARDOUS PETROLEUM AND SALT-CONTAMINATED SITES

by
George H. Holliday, PhD., P.E., BCEE
Holliday Environmental Services, Inc.
Bellaire, Texas
Lloyd E. Deuel, Jr., PhD., CPSSC
Whole Earth Environmental, Inc.
College Station, Texas

ASME
PRESS

Library of Congress Cataloging-in-Publication Data

Deuel, L. E.
   Guidebook for waste and soil remediation: for nonhazardous petroleum and salt-contaminated sites / Lloyd E. Deuel, Jr., George H. Holliday.
      p. cm.
   Includes bibliographical references and index.
   1. Soil remediation. 2. Oil pollution of soils. 3. Soils, Salts in. 4. Salt—Environmental aspects. I. Holliday, George H. II. Title.

TD879.P4D479 2008
628.5'5—dc22

2008047422

PREFACE

Although soil science has been applied to the production of agricultural crops for many years, few people apply soil science to remediation of nonhazardous oilfield wastes (NOW) or impacted soil found in industrial and governmental operations. We prepared this book to present to the industrial and governmental reader means of applying soil science technology to pit closures and impacted soil remediation problems. The book provides hands-on guidance for treating salt, pH, metals and hydrocarbon impacted soil and/or wastes. Also, the information is useful during pit closure so as to avoid post treatment problems and to comply with the various state rules. Also, the book contains necessary formulas, equations, time-tested criteria, and waste and soil amendment recommendations, so the reader can apply the demonstrated techniques to field problems. Recommended waste and soil analytical protocols are included in the book appendices. These protocols provide a means of verifying the procedures used by a laboratory during analyses are correct, accurate and repeatable.

We believe use of the material and recommendations contained in this book will leave a remediated site as good, if not better than, it was before operations commenced.

We express our gratitude to Shawn Hokanson, who assisted in preparing the figures and color plates found in the book.

George Holliday, Bellaire, Texas

Lloyd Deuel, Jr., College Station, Texas

CONTENTS

GLOSSARY OF
SOIL SCIENCE TERMS

1.1 TERMS

The following listing includes terms and definitions used in this book for describing soils and solid wastes.

Aeration, Soil—The process by which air in the soil is replaced by air from the atmosphere. In a well-aerated soil, the soil air is very similar in composition to the atmosphere above the soil. Poorly aerated soils usually contain a much higher percentage of carbon dioxide and a correspondingly lower percentage of oxygen than the atmosphere above the soil.

Air Porosity—The bulk volume of air in soil at any given time or under a given condition.

Angstrom (A°)—A measure of atomic structure or wavelength of light such that 1 A° equals 1×10^{-8} cm or 10 nanometer, nm (1×10^{-9} m).

Available Water—That portion of water in a soil readily absorbed by plant roots. Usually, available water is described as water held in the soil against a pressure between 1/3 and 15 atmospheres.

Bulk Density, Soil—The mass of dry soil per unit of bulk volume. The bulk volume is determined before drying.

Cation Exchange Capacity (CEC)—The total of the exchangeable cations a soil can adsorb, expressed in milliequivalents per 100 g (meq/100 g) or per gram (meq/g) of soil. The major cations in drilling waste solids or soil are calcium, magnesium, potassium and sodium.

Clay Mineral—(1) Naturally occurring inorganic crystalline material found in drilling wastes and soils and other earthen deposits, usually of clay size (<0.002 mm in diameter). (2) Material as described under (1), but not limited by particle size.

Clay Particle—(1) A waste solid or soil separate consisting of particles less than 0.002 mm in equivalent diameter. (2) Clay as a unit describes a textural class.

Crust—A surface layer of soil, ranging in thickness from a few millimeters to perhaps as much as an inch that is much more compact, hard and brittle when dry than the material immediately beneath it.

Deflocculate—(1) To separate compound particles into individual components by chemical and/or physical means. (2) To cause the particles of the dispersed phase of a colloidal system to become suspended in the dispersion medium.

Disperse—(1) To break up compound particles, such as aggregates, into the individual component particles. (2) To distribute or suspend fine particles, such as clay, in or throughout a dispersion medium, such as water.

Dry-Weight Percentage—The ratio of the weight of any one constituent of a soil to the oven-dry weight of the entire soil measured constant weight after drying at 105°C.

Electrical Conductivity (EC)—Conductivity measured directly in reciprocal units of resistance and reported in millimhos per centimeter (mmhos/cm) or millisiemens per centimeter (mS/cm). Electrical

Conductivity is an indirect measure of total dissolved solids (TDS). Electrial Conductivity can be determined from saturated paste, 1:1 by weight or 1:1 by volume extracts. Saturated paste electrical conductivity (SPEC) is preferred.

Exchange capacity—The total charge of the adsorption complex (sites) active in the adsorption of ions.

Exchangeable cation percentage—The extent to which the adsorption complex (sites) of a waste or soil is occupied by a particular cation.

Exchangeable sodium percentage (ESP)—The percentage of the cation exchange capacity of a soil occupied by sodium.

Fertility, soil—The status of a soil with respect to the amount and availability of elements necessary for plant growth.

Field capacity—Saturated soil water content available at a tension of 1/3 atmosphere.

Fine texture—Soil consisting of or containing large quantities of fine fractions, particularly of silt and clay.

Gravitational water—Water, which moves into, through or out of the soil under the influence of gravity.

Groundwater—That portion of the total precipitation which, at any particular time, is either passing through or standing in the soil or underlying strata; and is free to move under the influence of gravity.

Halophyte—A plant growing in salty soils; a plant thriving as opposed to tolerating a saline environment.

Heavy soil—A soil having a high content of fine separates, such as clay. Heavy soil is synonymous with fine-textured soil.

Humin—The fraction of the organic matter in a soil not dissolved upon extraction of the waste or soil with dilute alkali.

Humus—The more or less stable fraction of organic matter in the waste or soil remaining after the major portion of added plant and animal residues have decomposed.

Infiltration—The downward entry of water into a porous media such as earthen waste solids or soil.

Infiltration rate—A waste or soil characteristic determining or describing the maximum rate at which water can enter the waste or soil under specified conditions, including the presence of or an excess of water.

Kaolinite—(1) Aluminosilicate mineral of the 1:1 crystal lattice group, consisting of one silicon tetrahedral layer and one aluminum oxide-hydroxide octahedral layer. (2) The 1:1 group of the family of aluminosilicates. (Refer to Chapter 2 for discussion.)

Labile—Refers to a substance in soil readily transformed or presently available to plants.

Land classification—The arrangement of land units into various categories based upon the properties of the land or its suitability for some particular purpose, for example, farmland.

Land spreading—A process in which impacted wastes or soils are spread over a treatment area and tilled with native soil, without amendments.

Land treatment—A process in which impacted wastes or soils are spread over a treatment area and tilled with native soil. Nutrients, water and air are added to enhance biodegradation.

Landscape—The entirety of natural features such as fields, hills, forests, water, etc., that distinguish one part of the earth's surface from another part.

Lime, agricultural—A waste or soil amendment consisting principally of calcium carbonate, but including magnesium carbonate; used to furnish calcium and magnesium to neutralize soil acidity.

Macronutrient—A chemical element necessary in large quanities for the normal growth of plants and usually applied artificially in fertilizer or liming material.

Marsh—Periodically wet or continually flooded areas where the surface is not deeply submerged. Subclasses include freshwater and saltwater marshes.

Micronutrient—Chemical element necessary, but in extremely small amounts, for the normal growth of plants. Examples are: boron (B), chloride (Cl), copper (Cu), iron (Fe), manganese (Mn) and zinc (Zn).

Moisture tension (or pressure)—The equivalent negative pressure in the soil water. It is equal to the pressure applied to soil water to bring it to hydraulic equilibrium, through a permeable wall or membrane, with a pool of water of the same composition. The pressures used and the corresponding percentages most commonly determined are: (1) Fifteen-atmosphere percentage—the percentage of water contained in a saturated soil subjected to an applied pressure of 15 atmospheres until it is in equilibrium. (2) One-third-atmosphere percentage— the percentage of water contained in a saturated soil that has been subjected to an applied pressure of 1/3 atmosphere until it is in equilibrium.

Montmorillonite—An aluminosilicate clay mineral having a 2:1 expanding crystal lattice, for example, consisting of two silicon tetrahedral layers enclosing an aluminum octahedral layer. Considerable expansion may be caused by water moving between silica layers of contiguous units (interlayer water absorption). (See Chapter 2 for more discussion of 2:1 clays.)

Muck soil—(1) Soil containing between 20 and 50% organic matter. (2) An organic soil in which the organic matter is well decomposed.

Nitrogen (N)—Chemical element necessary for plant growth.

Nonsodic Waste/Soil—A waste or soil containing an exchangeable sodium percentage (ESP) ≤15%.

Oil—Petroleum hydrocarbons in crude or refined products.

Oven dry soil—Waste or soil dried at 105°C until it reaches constant weight.

Particle density—Mass per unit volume of the waste/soil particles.

Permeability, soil—The ease with which gases, liquids or plant roots penetrate or pass through a bulk mass or a layer of waste or soil.

Physical properties of wastes or soils—The characteristics, processes or reactions of a waste/soil, which are caused by physical forces and which can be described by an equation or expressed in physical terms.

Phosphorus (P)—Chemical element necessary for plant growth.

Porosity—Percentage of the total volume not occupied by waste or soil particles.

Potassium (K)—Chemical element necessary for plant growth.

Saline soil—A nonsodic waste or soil containing sufficient soluble salts to impair agronomic productivity. The electrical conductivity of the saturated paste extract is greater than 4 millimhos per centimeter (at 25°C) and the pH is usually less than 8.3 standard units

Saline sodic—(1) A waste or soil containing sufficient exchangeable sodium to interfere with the growth of most crop plants and containing appreciable quantities of soluble salts. (2) A waste or soil having an Exchangeable Sodium Percentage (ESP) >15% and the electrical conductivity (EC) of the saturation extract is >4 millimhos per centimeter.

Sand—A soil particle having an equivalent diameter of between 0.05 and 2.0 mm.

Saturation—(1) A condition in which all the voids between waste/soil particles are filled with liquid. (2) A condition in which the most concentrated solution possible is formed under a given set of physical conditions.

Saturated paste (SP) extract—A liquid sample representative of a saturated waste or soil.

Silt—A soil separate consisting of particles between 0.05 and 0.002 mm in equivalent diameter.

Sodic waste/soil—A waste or soil containing an Exchangeable Sodium Percentage (ESP) >15%.

Sodium adsorption ratio (SAR)—An empirical mathematical expression developed as an index to the sodium hazard in wastes and soils:

$$SAR = Na / [(Ca + Mg)/2]^{1/2}$$

High sodium (Na) levels (SAR > 12) in waste/soil solution cause calcium (Ca) and magnesium (Mg) deficiencies in plants.

Soil—(1) The unconsolidated mineral material on the immediate surface of the earth serving as natural medium for the growth of land plants. (2) The unconsolidated mineral matter on the surface of the earth having been subjected to and influenced by genetic and environmental factors of: parent material, climate, macro- and microorganisms and topography. These factors act over a period of time to produce a product-soil differing from the matter from which it is derived in many physical, chemical, biological and morphological properties and characteristics.

Solum—The upper and most weathered part of the soil profile.

Special management—Procedures used on pit or spill wastes requiring more complex closure procedures than simply mixing waste fluids with levee or background soil materials to complete closure.

Surface soil—Uppermost part of the soil, ordinarily moved in tillage operations.

Swamp—An area saturated with water throughout much of the year; usually the surface of the soil is not deeply submerged.

Tidal flats—Nearly flat areas, which are barren, and periodically covered by tidal waters.

Total dissolved solids (TDS)—The total concentration of dissolved cations (mg/L) in the waste or soil moisture.

Total petroleum hydrocarbons (TPH)—The total hydrocarbons (mg/kg) present in waste or soil as determined by a particular analytical method.

Wasteland—Land not suitable for, or capable of producing materials or services of value.

Wilting point—Water held in the soil against a pressure of 15 atmospheres.

1.2   SOIL AND PHYSICAL RELATIONSHIPS

The following conversion factors and equations provide a ready means of calculating quantities of soil amendments required for proper remediation.

Conversions
5.61 ft^3 = 1 barrel, bbl
Oil contains 78% carbon by weight
Carbon (mg/kg) = % TPH ×0.78 × 10,000 = C
1 ft^3 soil = 91.8 lb
1 acre = 43,560 ft^2

Equations

$$SAR = \frac{\text{Soluble Na (meq/L)}}{[\{\text{soluble Ca (meq/L)} + \text{soluble Mg (meq/L)}\}/2]^{1/2}}$$

$$ESP \% = [\text{exchangeable Na (meq/100g)}/\text{CEC (meq/100g)}] \times 100$$

Milliequivalents
Conversion to meq/L = cation (mg/L)/cation atomic weight
Cl(meq/L) = Cl(mg/L)/35.5(mg/meq)
Na(meq/L) = Na(mg/L)/23(mg/meq)
K(meq/L) = K(mg/L)/39(mg/meq)
Ca(meq/L) = Ca(mg/L)/20(mg/meq)
Mg(meq/L) = Mg(mg/L)/12(mg/meq)

Relationship of ESP to SAR

$$ESP = \frac{1.475 \times SAR}{1 + 0.0147 \times SAR}$$

EC Calculation

$$\text{Desired } EC_d = [(V_i \text{ impacted vol.}) \times (\text{measured } EC_i) + (V_b \text{ background vol.}) \times (\text{background } EC_b)]/(V_i \text{ impacted vol.} + V_b)$$

where

V_b (ft^3) = background soil volume needed for dilution, and
V_i (ft^3) = impacted waste or soil volume

Solving for V_b:

$$V_b = \frac{[(\text{measured EC}_i) - (\text{desired EC}_d)](V_i \text{ impacted volume})]}{(\text{desired EC}_d - \text{background EC}_b)}$$

Total volume, V_t (ft^3) = V_i impacted volume + V_b dilution volume

The above mass balance equation works for any contamination you have, such as TPH, SAR, ESP, etc.

SPEC = (1:1 by weight EC × 100)/SP moisture percent %

Where SPEC = saturated paste extract EC

Salt remediation gypsum and sulfur requirement determined from CEC and ESP

{(% ESP–12%)/100} × (CEC) = charge requirement (CR), in meq/100g

A CR of 1 meq/100 g = 1.7 tons gypsum (CaS0$_4$)/acre-ft waste or impacted soil, and

A CR of 1 meq/100 g = 0.3 tons elemental sulfur (S)/acre-ft waste or impacted soil.

Therefore: the gypsum requirement = CR × 1.7 (tons gypsum/acre-ft).

The sulfur requirement = CR × 0.3 (tons sulfur/acre-ft)

We recommend using *both* gypsum and sulfur when the contaminated materials are sodic (ESP > 15%), and *have solid phase calcium present* to react with sulfur. When using a combination of these materials; *the CR is divided by 2 to maintain the equivalent balance and avoid overtreatment.*

Relationship of API Gravity to Specific Gravity

$$\text{Sp. Gr.} = 141.5/(131.5 + \text{degree API})$$

CHAPTER

2

SOIL CHEMISTRY

2.1 COLLOIDS—THE REACTIVE SOIL FRACTION

Colloids are defined as soil separates smaller than 0.002 mm (2 µ) in particle size. They may either be mineral (inorganic) or organic in origin. Because of their small size, soil colloids have a large external surface area. Some layer silicates have an internal surface 50—80 times larger than the external surface and total surface area >800 m^2/g. The most important function of waste/soil colloids is the ability to attract and hold (adsorb) ions of opposite charge on the colloid particle surfaces. For most wastes/soil colloids, electronegative charges predominate, attracting positively charged ions or cations. The following discusses the source of electronegative charge in mineral and organic colloids. The influence of structural composition on properties of mineral colloids is also discussed in this chapter.

2.2 CLAY MINERALS

Clay minerals comprise the most significant group of colloids, which are collectively called mineral colloids. Chemical composition and atomic structure in combination define the various clay minerals by group or type. The atomic lattice of most clay minerals consists of two basic structural units: the *silica tetrahedral* and *aluminum octahedral sheets*. They are termed sheets, because of the leaflike or platy stacking.

2.2.1 Structural Units

The *silica tetrahedral sheet* is comprised of silicon atoms at the center of and equidistant from four oxygen atoms (Plate 1) arranged to form tetrahedrons (Plate 2), and joined at the base by sharing atoms in the same plane.

The *aluminum octahedral sheet* is comprised of aluminum atoms at the center of and equidistant from six oxygen or hydroxyl atoms (Plate 3). The octahedral arrangement of aluminum hydroxide $Al(OH)_3$ is demonstrated for kaolinite and gibbsite (Plate 4). Observe the close order packing of the octahedron units in both minerals and the planar arrangement of hydroxyl and oxygen atoms in the kaolinite structure.

Silicate clay minerals are classified according to the number and arrangement of tetrahedral and octahedral sheets. The *1:1 type minerals* consist of one silica tetrahedral sheet and one alumina octahedral sheet. The *2:1 type minerals* contain one alumina octahedral sheet sandwiched between two silica tetrahedral sheets.

2.2.1.1 **Type Clays.** Kaolinite is an example of 1:1 type clay. Plate 5 shows the chemical and structural features of a kaolinite. Notice the shared oxygen atoms holding the tetrahedral and octahedral sheets together. The clay layers are held together by hydrogen bonding, with no room for ions or small molecules between them (Plate 6). There is no interlayer space, which means

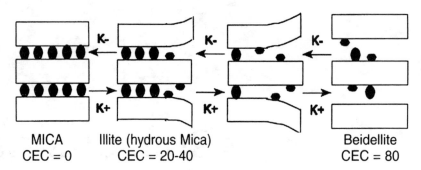

MICA
CEC = 0

Illite (hydrous Mica)
CEC = 20-40

Beidellite
CEC = 80

FIGURE 2.1
WEATHERING OF THE MINERAL AND LOSS OF POTASSIUM INCREASES CATION EXCHANGE CAPACITY (CEC)

colloidal properties are determined by external surfaces. Charges at the clay surface arise from broken edges and unsatisfied valences.

   2.2.1.2   Type Clays.   The 2:1 type clays consist of two silica tetrahedral layers and one alumina octahedral layer. The hydrous mica groups, of which illite is a member, carries an electronegative charge due to the substitution of aluminum (Al^{+3} valence cation) for silicon (Si^{+4}) in the tetrahedral sheet (Plate 7). The illite mineral has electronegative charges satisfied by potassium atoms (K^+), which are partially embedded in the tetrahedral sheet between clay layers. This holds the layers together and fixes the interlayer distance. The relatively small, fixed interlayer distance prevents water and cations from entering into the interlayer space (Plate 8). Weathering of the mineral and loss of potassium causes an increase in the effective electronegative charge, exposed interlayers, and increasing cation exchange capacity (Figure 2.1).

   The montmorillonite group differs from the hydrous mica group primarily in its expansible nature. In montmorillonite, the source of electronegative charge is the substitution of magnesium (Mg^{+2}) for aluminum (Al^{+3}) in the octahedral sheet (Plate 9). There is little attraction between oxygen atoms in adjacent tetrahedral layers, allowing the movement of water and free exchange of cations into the interlayer space (Plate 10).

TABLE 2.1
SOURCE OF PERMANENT CHARGE CATION EXCHANGE CAPACITY FOR MONTMORILLONITE

| | Substitution of Mg^{2+} for Al^{3+} | | | | |
| | Unsubstituted (Pyrophyllite) | | | Substituted Montmorillonite | |
	Ions	Charges +	Charges −	Ions	Charges +	Charges −
	$6 O^{-2}$		12	$6 O^{-2}$		12
Silica Layer	$4 Si^{+4}$	16		$4 Si^{+4}$	16	
	$4 O^{-2}, 2 OH^-$		10	$4 O^{-2}, 2 OH^-$		10
Alumina Layer	$4 Al^{+3}$	12		$4 Al^{+3}, 2 Mg^{+2}$	10	
	$4 O^{-2}, 2 OH^-$		10	$4 O^{-2}, 2 OH^-$		10
Silica Layer	$4 Si^{+4}$	16		$4 Si^{+4}$	16	
	$6 O^{-2}$		12	$6 O^{-2}$		12
	Total	44	44		42	44
	Net		0			2(−)

TABLE 2.2
MINERAL SOURCES OF pH-DEPENDENT CHARGES

Bonding	Designation	Example	pKa (Acid)
Al‒1/2 / Al‒OH‒1/2	Aluminohydroxyl	$Al(OH)_3$	12+
Si‒1‒OH	Silica Broken Edge	H_2SiO_3	9.5
Si / Al‒1/2‒OH$^{1/2+}$	Positive Broken Edge	Si / Al‒OH$^{1/2+}$	7.0
Al$^{1/2}$ 3/4‒O‒3/4 ‒H$^{1/4+}$ / H$^{1/4+}$	Substituted Hydronium	$Al(OH_2^{1/2+})_6$	5.0

2.3 MINERAL EXPANSIBILITY

Cations, such as magnesium (Mg^{+2}) and calcium (Ca^{+2}), are hydrated by six water molecules (Plate 11). This maintains an interlayer spacing of approximately 12 angstroms (A°) for montmorillonite clay. Due to large hydrated radii, the interlayer distance for sodium (Na^+) and lithium (Li^+) saturated (predominately adsorbed) montmorillonite, is not bounded by the cohesive forces exhibited by divalent cations. This results in particle separation and dispersion. It is this property of montmorillonites, which make it useful as a drilling fluid. However, this beneficial attribute in drilling fluids must be overcome for successful land treatment of the drilling waste during remediation.

2.4 SOURCES OF CEC

CEC is the total quantity of exchangeable cations a waste or soil can adsorb, and is expressed in milliequivalents of positive charge per 100 g of waste or soil (meq/100 g). Permanent charge is a term defining the charge generated from isomorphic substitution as demonstrated in the charge balance presented in Table 2.1. Table 2.1 summarizes the sources of net negative charges for montmorillonite created by Mg^{+2} substituting for Al^{+3} in the aluminum layer. Other mineral sources of CEC are pH-dependent and may be attributed to hydroxylation and removal of a proton resulting in a net nega-

TABLE 2.3
ORGANIC SOURCES OF pH-DEPENDENT CHARGES

Functional Group	Structural Formula	pKa (Acid)
Normal phenol	⬡‒OH	6 – >10
Acid phenol	NO_3 / NO_3‒⬡‒OH / NO_3	<4
Normal carboxyl	R‒C‒OH	~ 4 – >6
Acid carboxyl	Cl / R‒C‒C‒OH / Cl	<3

tive charge (Table 2.2). Similar functional groups associated with organic molecules, also result in pH-dependent sources of negative charges (Table 2.3). Figure 2.2 demonstrates how pH dependence can affect the net charge of various clays.

The cation exchange capacity of soils varies according to pH. These differences are caused by variations in mineralogy, clay content and organic matter (Figure 2.3). Table 2.4 shows the CEC of common soil materials. The high CEC of organic matter makes organic matter particularly effective for treating salt impacted wastes and soils. Table 2.5 illustrates the effect of soil texture on CEC. Generally, soils with high CEC are more fertile than soils having low CEC. Also, the finer the soil texture the higher the CEC.

2.5   PHYSICAL PROPERTIES OF SOILS

2.5.1   Soil Texture

Soil texture is the size distribution of primary, inorganic particles in soils. The particle sizes range from sand (2–0.05mm), silt (0.05–0.002 mm), to clay (<0.002 mm). The surface area of clay is 100–10,000 times greater than sand. A *loam* is a combination of the three particle sizes exhibiting the properties of sand, silt and clay.

2.5.2   Particle and Bulk Density

Particle density is density of the waste/soil particles. Bulk density is a measure of the weight per unit volume (g/cm^3) of waste or soil. Sandy soils generally have greater bulk density than well-aggregated

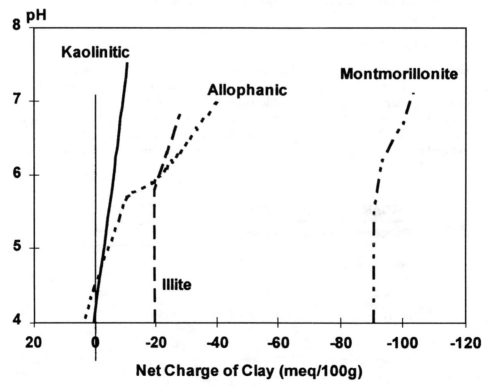

FIGURE 2.2
NET CHARGES OF SELECTED CLAYS AT VARIOUS pH LEVELS

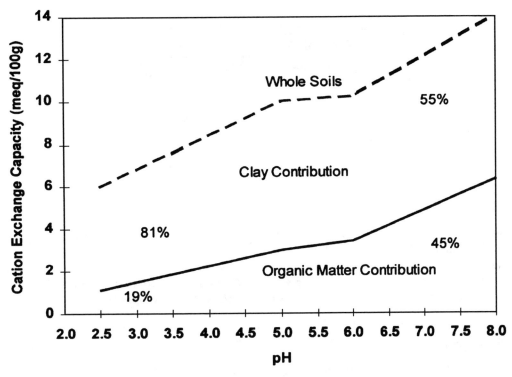

FIGURE 2.3
CEC OF SOIL AND SOIL COMPONENTS ACCORDING TO pH

finer-textured soils (Table 2.6). Percent pore space is the space available within the waste or soil for air or water. It is calculated as:

$$\% \text{ Pore space} = 100 - [(\text{bulk density/particle density}) \times 100]$$

Soils having greater bulk density have lower percentage porosity (Table 2.7). Macropores are larger pores allowing the ready movement of air and percolation of water. Micropores are small pores in which air movement is impeded and water moves slowly due to capillary rise. A sandy soil has a lower total porosity than clay, but air and water movement is rapid because of dominance of macropores. In clays, water and air movement is slow even though porosity is greater than sands, because of micro-

TABLE 2.4
CEC OF COMMON EXCHANGE MATERIALS

Material	CEC, mEq/100 g
Organic Matter	100–300
Vermiculite	100–150
Montmorillonite	60–100
Chlorite	20–40
Illite	20–40
Kaolinite	2–16
Sesquioxides	0

TABLE 2.5
RELATIONSHIP BETWEEN WASTE
OR SOIL TEXTURE AND CEC

Textural Class	CEC, mEq/100 g
Sand	1–5
Sandy loam	5–10
Loam	5–15
Silt loam	5–15
Clay loam	15–20
Clay	>30

pore dominance. Water infiltration is faster in loamy sand than clay loam. For a soil with a clay loam over sand, the water infiltrates the clay loam layer and does not enter the sand layer until the clay loam layer is saturated because of the abrupt textural change between layers (Figure 2.4).

Soil aeration is important for maintaining waste and soil under aerobic conditions. Aeration is affected by texture, structure, total porosity and macropores versus micropores.

2.5.3   Soil Structure

Soil structure is the way the soil particles group together into stable forms or aggregates. Common types of soil structure are granular, single grain; prismatic, blocky, massive and platy (Figure 2.5). Soil structure affects water infiltration. Soils with a granular or single-grain structure can have rapid infiltration, for example, water moves quickly through these soils. Prismatic and blocky soils have moderate infiltration. Massive and platy soils have slow infiltration, because vertical movement is limited, since the water must move downward through the micropores in the soil.

Soil aggregation and structure are influenced by many factors. Aggregation tends to increase under conditions of

- Wetting/drying
- Freezing/thawing
- Physical activity of roots and animals
- Decaying organic matter/microbes

Tillage increases erosion and can affect aggregation negatively due to the formation of a hardened plow pan beneath the surface. The effect of adsorbed cations depends on the cations; for example, *calcium improves aggregation, while large amounts of sodium destroy structure.* Sodium saturated soils become structureless soils and are often hard-packed requiring mechanical treatment (chiseling or plowing) to permit them to receive calcium amendments, air and water during land treatment.

TABLE 2.6
SOIL TEXTURE EFFECT ON BULK DENSITY

Textural Class	Bulk Density, g/cm³
Sands and sandy loams	1.2–1.8
Loams, silt loams, clay loams	1.3–1.6
Clay	1.0–1.5

TABLE 2.7
RELATIONSHIP BETWEEN SOIL TEXTURE, BULK DENSITY AND POROSITY

Textural Class	Bulk Density, g/cm^3	Porosity, %
Sand	1.55	42
Sandy loam	1.40	48
Fine sandy loam	1.30	51
Loam	1.20	55
Silt loam	1.15	56
Clay loam	1.10	59
Clay	1.05	60
Aggregated clay	1.00	62

2.5.4 Organic Matter

Soil organic matter is very important for aggregation and structure. Other soil constituents increase aggregation, too. In order of importance, these are

- Microbial gum
- Iron oxide
- Organic carbon
- Clay

Soil organic matter increases aggregation, water holding capacity, nutrient reserve, and lowers the bulk density. Decreased bulk density results in increased porosity and, when combined with improved structure, results in increased macroporosity. This opens the soil and allows for the infiltration and

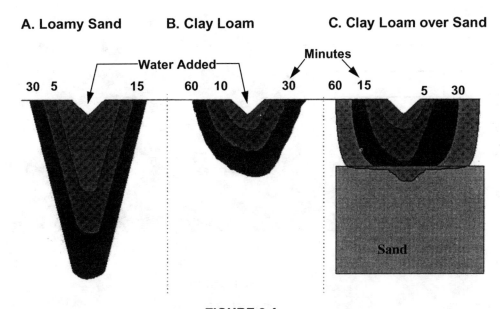

FIGURE 2.4
WATER INFILTRATION PROFILE FOR CLAY LOAM WITH AND WITHOUT SAND UNDERLAYER

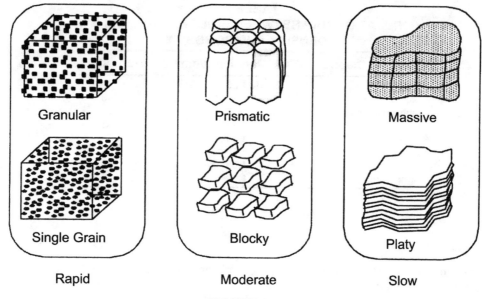

FIGURE 2.5
COMMON SOIL STRUCTURES

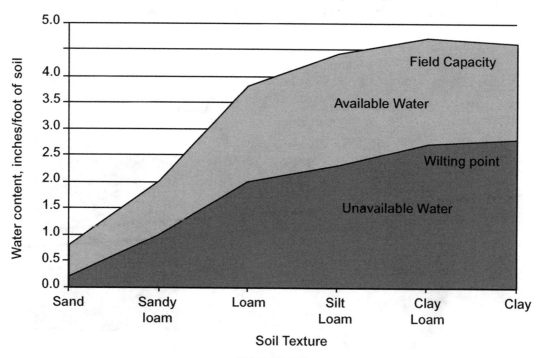

FIGURE 2.6
AVAILABLE WATER AS INFLUENCED BY SOIL TEXTURE

TABLE 2.8
AVAILABLE WATER CAPACITIES OF
VARIOUS SOIL TEXTURES

Textural Class	Water per Foot of Soil, in.
Medium sand	0.9
Fine sand	1.1
Sandy loam	1.4
Fine sandy loam	1.8
Loam	2.0
Silt loam	2.1
Clay loam	2.0
Clay	1.4

percolation of rainfall or irrigation water necessary to ameliorate saline and/or sodic soils. (Refer to Appendix I for the impact of excess salt and sodicity in soils.)

Organic matter is an important source of CEC in soils. *Adding organic matter increases the ability of waste or soils to adsorb cations and therefore serves as a sink for excess salt.*

2.6   SOIL WATER CONTENT AND AVAILABILITY

There are several commonly defined levels of soil water content. *Field capacity* is the water content held in the soil after drainage due to the pull of gravity and matrix effects. The *wilting point* is defined as the water content at which plants can no longer extract water from the soil. The *available water* content of a soil is defined as the quantity of water held between the field capacity and the wilting point (Figure 2.6).

Soil texture has a great effect on both the field capacity and wilting point of soil. Coarse-textured soils (sands and sandy loams) have much lower field capacities and wilting points than fine-textured soils (clays) (Table 2.8 and Figure 2.6). The net effect is the coarse-textured soils contain much less available water and are more susceptible to salt contamination. On the other hand, excess salts are more readily managed in course textures soils by leaching technologies. The addition of organic matter to either coarse- or fine-textured soils or waste solids aid in reducing the level of salt in soil solution.

Appendix V provides more information on the relationship between field moisture equivalents, saturated soil conditions and porosity.

CONTAMINATION AND IMPACTS OF EXPLORATION AND PRODUCTION WASTE CONSTITUENTS

3.1 INTRODUCTION

Impacted waste and soil result from (1) drilling operations where drilling fluids become contaminated from discharging engine oil, chemicals or waste into the reserve pit, (2) producing operations where oil, condensate, chemicals or produced water are spilled or released onto the ground or into the reserve pit, and (3) pipeline breaks or leaks where oil and/or produced water are released to the soil.

3.2 DRILLING OPERATIONS

Rotary drilling operations result in generating drilling fluids. Drilling fluids can be:

1. Fresh water base
2. Saltwater base
3. Oil base
4. Synthetic base
5. Emulsion muds

Regardless of the type of muds used, the same contaminants frequently are observed, including changes in pH, Electrical Conductivity, Sodium Adsorption Ratio, Cation Exchange Capacity, Exchangeable Sodium Percentage and total metals. *Pit solids exhibit an increased level of some constituents (up to 30 times more than the pit liquids) with an attendant increased potential for an adverse impact on the environment.*

3.2.1 pH

Drilling fluids typically are alkaline (pH > 10) during use. This high pH disperses the clay and increases the effectiveness of the drilling fluid. The high pH results from the addition of lye, soda ash, or other caustics. Without continued maintenance of pH, natural aging and weathering of the drilling fluid causes a decrease of pH through absorption of atmospheric CO_2. The desired agricultural pH range is 5.5–8.3 standard units (s.u.). However, within the United States the mandated range of pH is 6–9 s.u., as established from discharge criteria from water discharge permits.

TABLE 3.1
FREQUENCY DISTRIBUTION FOR pH IN DRILLING RESERVE PIT SOLIDS

Concentration Interval (s.u.)	Frequency Distribution	Percent of Total	Cumulative Percent
<2.0	2	0.13	0.13
2.1–4.0	10	0.64	0.77
4.1–6.0	140	8.86	9.73
6.1–9.0	1092	69.91	79.64
9.1–11.0	265	16.97	96.61
>11.0	53	3.39	100.0
Total	1562		
Maximum	12.6		
Minimum	1.7		
Average	7.92		

pH determinations for 1,562 E&P reserve pit samples averaged 7.92 s.u., ranging from pH 1.7 to pH 12.6 (Table 3.1). Cumulative frequency distributions demonstrated 79.6% of the pits fall within the pH 6–9 range, with 9.7% less than pH 6 and 20.4% greater than pH 9.

3.2.2   Electrical Conductivity (EC)

Electrical conductivity (EC) is the common measure of soil salinity. EC is critical to E&P waste and impacted soil characterization because of the potential for high brine content and the adverse impact brine has on plant growth and water quality. Produced water and pit liquids are analyzed for EC directly and reported in mmhos/cm. EC is measured for soils and solid wastes using a deionized water extract at a 1:1 (soil:water by weight) or saturated paste (SP) moisture equivalent (Appendix I, Annex B).

Salt distributions measured for 1603 reserve pit samples ranged in value from EC 0.2 mmhos/cm to 1559 mmhos/cm with an average EC 34.8 mmhos/cm (Table 3.2). Cumulative percentages show problematic salt levels in >68% of pits surveyed. Problematic is defined as materials requiring some management at closure (EC > 8.0 mmhos/cm). Acceptable ranges are controlled by post-closure land use as discussed in Chapter 7 and Appendix I.

TABLE 3.2
FREQUENCY DISTRIBUTION FOR SATURATED PASTE EC IN DRILLING RESERVE PIT SOLIDS

Concentration Interval (mS/cm)	Frequency Distribution	Percent of Total	Cumulative Percent
<4.0	341	21.3	21.3
4.1–8.0	170	10.6	31.9
8.1–25.0	643	40.1	72.0
25.1–50.0	229	14.3	86.3
50.1–100	101	6.3	92.6
>100	119	7.4	100.0
Total	1603		
Maximum	1559		
Minimum	0.2		
Average	34.8		

3.2.3 Sodium Adsorption Ratio (SAR)

Soluble cationic distributions are needed to assess the potential sodium damage from wastes. The sodium adsorption ratio (SAR) is related to poor soil physical characteristics and plant toxicities. It is defined by an empirical function of sodium, calcium and magnesium. SAR is computed as follows:

$$SAR = Na/[(Ca + Mg)/2]^{1/2}$$

where ationic concentrations are expressed in milliequivalents per liter (meq/L), and determined from a saturated paste extract. The soil saturated paste moisture content (SP % Moi) is defined as the maximum amount of water held in the puddled soil without free water collecting in a depression made in the soil mass.

SAR is used in conjunction with SPEC to evaluate the potential damages associated with sodium salts. SAR distributions determined for 1,464 reserve pit solids ranged from SAR less than 0.1 to 2661 with an average SAR of 7.92 (Table 3.3). About 62% of the pit solids analyzed, as well as the average SAR, fail the criteria of SAR<12 suggested in the Limiting Constituents for Land Disposal of Petroleum Extraction Industry Wastes (Chapter 7 and Appendix I).

3.2.4 Cation Exchange Capacity (CEC)

The individual clay particles in soils and drilling fluids have a negatively charged surface on which different positively charged cations are adsorbed. The sum of the positive charges of the adsorbed cations is equal to the sum of the negative surface charge or cation exchange capacity (CEC). Replacement of adsorbed cations by a chemically equivalent number of cations dissolved in soil solution occurs readily. The reaction is reversible depending upon the equilibrium between soluble and adsorbed (exchangeable) cations.

CEC criterion for drilling fluids is given as a guide for evaluating acceptable metal loadings. CEC criterion is not a point for consideration when muds contain no appreciable metals. Drilling fluids will generally have much larger CEC values than existing surface soils. In addition, the CEC measurement is required to estimate the waste or soil Exchangeable Sodium Percentage (ESP) value. We measure CEC in meq/100 g. The CEC distribution for 1455 pit solids ranges from a low of less than 1 meq/100 g to a high of 84.7 meq/100 g and averages 22.3 meq/100 g (Table 3.4).

TABLE 3.3
FREQUENCY DISTRIBUTION FOR SODIUM ADSORPTION
RATIO (SAR) IN DRILLING RESERVE PIT SOLIDS

Concentration Interval (unitless)	Frequency Distribution	Percent of Total	Cumulative Percent
<12.0	462	31.6	31.6
12.1–14.0	91	6.2	37.8
14.1–25.0	286	19.5	57.3
25.1–50.0	309	21.1	78.4
50.1–00	185	12.6	91.0
>100	131	9.0	100.0
Total	1464		
Maximum	2661		
Minimum	0.1		
Average	34.1		

TABLE 3.4
FREQUENCY DISTRIBUTION FOR CATION EXCHANGE
CAPACITY (CEC) IN DRILLING RESERVE PIT SOLIDS

Concentration Interval (mEq/100g)	Frequency Distribution	Percent of Total	Cumulative Percent
<7.0	128	8.8	8.8
7.1–14.0	249	17.1	25.9
14.1–21.0	288	19.8	45.7
21.1–28.0	392	26.9	72.6
28.1–35.0	256	17.6	90.2
>35.0	142	9.8	100.0
Total	1455		
Maximum	84.7		
Minimum	0.8		
Average	22.3		

3.2.5   Exchangeable Sodium Percentage (ESP)

Just as salt-affected muds are characterized and classified by their content of soluble salts, sodic (excess Na) muds are characterized by the exchangeable sodium percentage (ESP > 15%, Appendix I). Saline-sodic drilling fluids contain excess salt as well as excess sodium (SPEC > 8 mmhos/cm *and* ESP>15%).

When exchangeable sodium is in excess, there is a general lack of structural stability among soil particles, and water infiltration is impeded. Salinity hazards coupled with sodic conditions create soil remediation problems due to the inherently slow infiltration and percolation available to move excess salts out of the root zone.

Sodic and saline-sodic drilling fluids in sensitive areas (such as farmland) are not easily managed. Reduced treatment thickness is required for drilling fluids having a high (>15%) ESP. Soil amendments, such as sulfur, to adjust associated alkalinities, and calcium amendments to desorb sodium are often required in sensitive areas.

The frequency distribution for 1455 pits shows ESP values ranging from <0.1 to 100%, with an average ESP of 26% (Table 3.5). More than 57% of the pit solids analyzed fail the criteria of ESP = 15% suggested in Chapter 7 and Appendix I.

TABLE 3.5
FREQUENCY DISTRIBUTION FOR
EXCHANGEABLE SODIUM PERCENTAGE (ESP)
IN DRILLING RESERVE PIT SOLIDS

Concentration Interval (percent)	Frequency Distribution	Percent of Total	Cumulative Percent
<15.0	613	42.1	42.1
15.1–25.0	277	19.0	61.1
25.1–35.0	196	13.5	74.6
35.1–50.0	167	11.5	86.1
50.1–75.0	115	7.9	94.0
75.1–100	87	6.0	100.0
Total	1455		
Maximum	100		
Minimum	<0.1		
Average	26.0		

3.2.6   Total Metals

The chemistry of metals in soil or solid waste matrices is exceedingly complex and as such not completely understood. The ultimate fate of metals contained within a given matrix is controlled by solvation, complexing, chemisorption, and cation exchange processes. These processes control bio-availability and mobility within the receiving soil matrix. Heavy metals are labile in soils and wastes.

Metals are distributed in various forms or fractions comprising the solid phase matrix. This includes availability through water soluble, exchangeable, bound-iron/manganese oxides, bound-to carbonates, bound-to organics, and that entrained as part of the mineral fabric. The labile pool generally includes the water soluble, exchangeable, and a portion of the organic fraction under moderately well aerated and/or oxidizing conditions. A portion of the heavy metals bound to iron/manganese oxides also may become a significant source of labile metals under poorly aerated and/or reducing conditions.

Considerable concerns are raised regarding metal in drilling fluids and cutting. Accordingly, the authors analyzed over a thousand pits to establish the extent of the perceived problem. We observed few cases where high concentrations of metals result from drilling operations. *Water-soluble and exchangeable* forms of metals define the principal risk associated with heavy metals in the natural environment. If the metal is not in one of these forms, it poses no significant threat or hazard to the environment. Water-soluble and exchangeable metal fractions generally are zero or in trace quantities for most E&P wastes and are of no concern in management design. Unfortunately, the natural environment cannot be controlled; redox or pH conditions may change, potentially releasing a measurable and significant fraction of a given heavy metal.

It is our opinion; only *total metals* provide a measurable index for assessing E&P wastes relative to environmental impact and management design. We recommend E&P wastes routinely be analyzed for true total barium, chromium, lead and zinc. Total arsenic, cadmium, mercury, selenium and silver should be included when characterizing a field for the first time. However, experience indicates metal contaminants seldom are found.

Research shows *total metals released under EPA SW-846, Method 3050 protocol* provide a reasonable index for all metals except *barium*. Barium is best analyzed under the protocol given in the Laboratory Procedures Manual for the Analysis of Oilfield Waste (Louisiana Department of Natural Resources, August 9, 1988 or latest revision). The limiting criterion (Chapter 7) for barium established for drilling fluids (20,000–40,000 mg/kg) are based on total digestion following Louisiana protocol, however where soil-waste mixes are buried, we recommend a true total barium concentration not to exceed 100,000 mg/kg.

3.2.6.1   Silver (Ag).   Silver is included in the list of total metals typically screened for in E&P wastes, because it is one of the Resource Conservation and Recovery Act (RCRA) metals. It forms

TABLE 3.6
FREQUENCY DISTRIBUTION FOR SILVER (Ag)
IN DRILLING RESERVE PIT SOLIDS

Concentration Interval (ppm)	Frequency Distribution	Percent of Total	Cumulative Percent
<1.0	707	66.2	66.2
1.1–3.0	246	23.0	89.2
3.1–5.0	65	6.1	95.3
5.1–7.0	24	2.2	97.5
7.1–9.0	16	1.5	99.0
>9.0	10	1.0	100.0
Total	1068		
Maximum	46.4		
Minimum	<0.1		
Average	1.6		

TABLE 3.7
FREQUENCY DISTRIBUTION FOR ARSENIC (As)
IN DRILLING RESERVE PIT SOLIDS

Concentration Interval (ppm)	Frequency Distribution	Percent of Total	Cumulative Percent
<2.0	465	44.9	44.9
2.1–5.0	276	26.7	71.6
5.1–7.0	91	8.8	80.4
7.1–10.0	83	8.0	88.4
10.1–20.0	75	7.2	95.6
>20.0	45	4.4	100.0
Total	1035		
Maximum	122.2		
Minimum	<0.1		
Average	4.21		

precipitates in soils with chlorides, sulfates and carbonates and therefore is unlikely to enter the food chain by plant uptake. Silver is allowed in drilling mud pit soil mixtures at levels below 200 mg/kg (Chapter 7). Of 1,068 pits tested, 99% contained silver levels <10 mg/kg (Table 3.6).

3.2.6.2   Arsenic (As).   Arsenic naturally occurs in soil in quantities ranging from 1 to 40 mg/kg. We recommend arsenic be =10 mg/kg for waste-soil mixtures. Eighty-eight percent of 1035 pits tested for arsenic under =10 mg/kg (Table 3.7). Under anaerobic conditions arsenic can be reduced to a form, which is toxic to plants. There is still some debate whether arsenic is an essential element for plant growth.

3.2.6.3   Barium (Ba).   In soils, the barium content ranges from 100 to 3000 mg/kg. Barite ($BaSO_4$) is commonly used as an additive to the drilling fluids to increase the mud weight. Barite is insoluble in water and the mineral acids used in the EPA total metals digest procedure. Frequency distributions for 1,129 E&P waste pit samples using EPA test protocol showed 92.2% of the pits tested <8000 mg/kg barium (Table 3.8). We find the EPA protocol provides inaccurate and inconsistent results for barium analysis.

3.2.6.4   True Total Barium (TTBa).   A more accurate level of barium is measured using a special procedure from the Louisiana recommended method (Rule 29-B, 2000), which measures the true total

TABLE 3.8
FREQUENCY DISTRIBUTION FOR BARIUM (Ba)
IN DRILLING RESERVE PIT SOLIDS

Concentration Interval (ppm)	Frequency Distribution	Percent of Total	Cumulative Percent
<500	459	40.7	40.7
501–1,000	210	18.6	59.3
1,001–2,000	115	10.2	69.5
2001–4,000	127	11.2	80.7
4,001–8,000	130	11.5	92.2
>8,000	88	7.8	100.0
Total	1,29		
Maximum	91,100		
Minimum	<10		
Average	2,308		

TABLE 3.9
FREQUENCY DISTRIBUTION FOR TRUE TOTAL
BARIUM (TTBa) IN DRILLING RESERVE PIT SOLIDS

Concentration Interval (ppm)	Frequency Distribution	Percent of Total	Cumulative Percent
<5,000	599	48.8	48.8
5,001–10,000	141	11.5	60.3
10,001–20,000	74	6.0	66.3
20,001–40,000	44	3.6	69.9
40,001–80,000	126	10.3	80.2
>80,000	243	19.8	100.0
Total	1,227		
Maximum	210,841		
Minimum	<10		
Average	31,535		

barium. Frequency distribution for true total barium analyzed for 1,227 E&P waste pits using Louisiana 29-B protocol showed 66.3% of the pits are below the wetland regulatory level of =20,000 mg/kg and 69.9% of the pits wastes are acceptable under the upland level of 40,000 mg/kg (Table 3.9). We recommend *true total barium* be maintained =20,000 mg/kg for wetlands and =40,000 mg/kg for upland waste-soil mixtures and 100,000 mg/kg for burial (as previously discussed under Total Metals).

3.2.6.5 Cadmium (Cd).   Depending on the salinity, chemical hardness and pH, cadmium can be found in reserve pits in the form of CdO, $Cd(OH)_2$ CdS and $CdCO_3$. Cadmium concentrations as low as 1 mg/kg in soil can reduce plant growth in certain species. We recommend maintaining cadmium =10 mg/kg in waste-soil mixtures. Of the 1028 pit solids analyzed 99% were below the recommended level of 10 mg/kg (Table 3.10).

3.2.6.6 Chromium (Cr).   Chromium naturally occurring in soils ranges from 5 to 3000 mg/kg, with an average level of about 100 mg/kg. Chromium is not very soluble under the alkaline conditions associated with most drilling fluids. Of the 1.503 pit solids tested, 98.7% of them were =500 mg/kg chromium (Table 3.11). We recommend chromium concentrations be maintained =500 mg/kg in waste-soil mixtures.

3.2.6.7 Mercury (Hg).   Mercury naturally occurs in soils in concentrations of <0.2 mg/kg. Mercury often forms stable complexes with clays and organic matter. Only 2 of the 1005 pits tested showed

TABLE 3.10
FREQUENCY DISTRIBUTION FOR CADMIUM (Cd)
IN DRILLING RESERVE PIT SOLIDS

Concentration Interval (ppm)	Frequency Distribution	Percent of Total	Cumulative Percent
<2.0	911	88.6	88.6
2.1–5.0	85	8.3	96.9
5.1–10.0	22	2.1	99.0
10.1–0.0	6	0.6	99.6
20.1–30.0	3	0.3	99.9
>30.0	1	0.1	100.0
Total	1028		
Maximum	37.0		
Minimum	<0.1		
Average	1.04		

TABLE 3.11
FREQUENCY DISTRIBUTION FOR CHROMIUM (Cr)
IN DRILLING RESERVE PIT SOLIDS

Concentration Interval (ppm)	Frequency Distribution	Percent of Total	Cumulative Percent
<50	1060	70.5	70.5
51–250	398	26.5	97.0
251–500	25	1.7	98.7
501–1000	7	0.4	99.1
1,001–2500	5	0.3	99.4
>2500	8	0.6	100.0
Total	1503		
Maximum	11,300		
Minimum	<1		
Average	81.5		

mercury levels >10 mg/kg; that is, 99.7% of the pits solids were <10 mg/kg (Table 3.12). We recommend maintaining mercury levels in waste-soil mixtures at or below 10 mg/kg.

3.2.6.8　Lead (Pb).　For native soils, lead concentrations are usually about 15 mg/kg. Lead is normally found in insoluble forms in drilling muds. When high chloride contents are present in drilling muds, soluble lead forms may exist. We recommend 500 mg/kg as the maximum lead allowed in waste-soil mixtures. Of the 1536 pits tested 97.6% were at or below this level (Table 3.13).

3.2.6.9　Selenium (Se).　Selenium occurs naturally in U.S. soils in the range of 1–7 mg/kg. Selenium usually occurs in the forms of selenate and selenite, which are soluble. Selenium is an essential element for animals, but not for plants. Toxic levels of selenium in soil can cause stunting and chlorosis in plants. We recommend 10 mg/kg maximum selenium in waste-soil in mixtures. Nine hundred and fourteen pits were analyzed for selenium, and 99.7% of these pits were below the level of 10 mg/kg (Table 3.14).

3.2.6.10　Zinc (Zn).　Zinc levels in soils naturally range from 10 to 300 mg/kg. Zinc is an essential plant element, needed for the formation of plant hormones and protein synthesis. Zinc is phytotoxic at levels above 600 mg/kg. We recommend an allowable level of =500 mg/kg zinc in waste-soil mixtures. Zinc was analyzed in 1535 pit samples, 83.6% of the drilling mud pits contain =500 mg/kg zinc (Table 3.15).

TABLE 3.12
FREQUENCY DISTRIBUTION FOR MERCURY (Hg)
IN DRILLING RESERVE PIT SOLIDS

Concentration Interval (ppm)	Frequency Distribution	Percent of Total	Cumulative Percent
<1.0	869	86.4	86.4
1.1–2.0	96	9.5	95.9
2.1–5.0	27	2.6	98.5
5.1–0.0	13	1.3	99.8
>10.0	2	0.2	100.0
Total	1005		
Maximum	23.8		
Minimum	<0.1		
Average	0.58		

TABLE 3.13
FREQUENCY DISTRIBUTION FOR LEAD (Pb)
IN DRILLING RESERVE PIT SOLIDS

Concentration Interval (ppm)	Frequency Distribution	Percent of Total	Cumulative Percent
<50	854	55.6	55.6
51–250	599	39.0	94.6
251–500	46	3.0	97.6
501–750	19	1.3	98.9
751–1,000	5	0.3	99.2
>1,000	13	0.8	100.0
Total	1536		
Maximum	3546		
Minimum	<1		
Average	87.8		

TABLE 3.14
FREQUENCY DISTRIBUTION FOR SELENIUM (Se)
IN DRILLING RESERVE PIT SOLIDS

Concentration Interval (ppm)	Frequency Distribution	Percent of Total	Cumulative Percent
<1.0	825	90.3	90.3
1.1–4.0	76	8.3	98.6
4.1–8.0	9	1.0	99.6
8.1–10.0	1	0.1	99.7
10.1–15.0	2	0.2	99.9
>15.0	1	0.1	100.0
Total	914		
Maximum	18.0		
Minimum	<0.1		

TABLE 3.15
FREQUENCY DISTRIBUTION FOR ZINC (Zn)
IN DRILLING RESERVE PIT SOLIDS

Concentration Interval (ppm)	Frequency Distribution	Percent of Total	Cumulative Percent
<100	678	44.2	44.2
101–250	336	21.9	66.1
251–500	268	17.5	83.6
501–1000	195	12.7	96.3
1001–4,500	55	3.6	99.9
>2,500	3	0.1	100.0
Total	1535		
Maximum	6650		
Minimum	<1		
Average	278		

3.2.7 Petroleum Hydrocarbon

Crude oil from a producing formation, and diesel or mineral oils added to drilling muds are the chief sources of petroleum hydrocarbons in E&P drilling wastes. Of these, diesel is the most troublesome with respect to waste management. Fortunately, diesel responds well to bioremediation.

States regulating E&P waste disposal vary in test parameters and applicable treatment thresholds (Chapter 7). Most states use either *oil and grease* (O&G) by gravimetric assay or total petroleum hydrocarbon (TPH) by *infrared spectroscopy* (IR) for determination of acceptance values. Analytically, the methods yield similar results for unweathered oils. O&G levels tend to be slightly lower than TPH for fresh oils in which the lighter hydrocarbons have not been dissipated. Additionally, total organic carbon (TOC) can be used as an alternative analytical method for measuring petroleum hydrocarbons. In our opinion, the best analytical method for TPH waste or soil is *gas chromatography with flame ionization detection* (GC-FID). This method accurately follows the conversion of TPH to humus during bioremediation. The following discussion is limited to those methods used by the states.

3.2.7.1 Oil & Grease. Oil and grease determined by solvent extraction and gravimetric analysis for 1584 reserve pits averaged 4.3% O&G (43,000 mg/kg), ranging from a low of <0.1% to a high of almost neat oil at 99.1% (Table 3.16). Some 36.3% of the pits in this database meet the Louisiana closure threshold of =1% O&G. Although the data was not split out at the Oklahoma threshold of 2% O&G, it is reasonable to believe >50% of the pits would pass this standard prior to any treatment. We recommend levels in waste-soil mixtures at or below 1% TPH (Chapter 7).

3.2.7.2 Total Petroleum Hydrocarbon—Infrared (TPH-IR). Some regulatory bodies require TPH-IR analyses for inventory purposes and comparison to negotiated treatment standards. TPH data for 139 E&P reserve pits receiving only drilling muds is summarized in Table 3.17. The average was 62,263 mg/kg TPH (6.2%) with a minimum at 13 mg/kg TPH and maximum at 519,336 mg/kg TPH. Cumulative percentage of pits passing 10,000 mg/kg TPH (1% TPH) was 41.7%, and pits =50,000 mg/kg TPH (5%) was 65.4%. These values correspond well to the cumulative O&G percentages previously discussed (36.4% = 1%, and 69.2 = 5%).

3.2.7.3 Total Organic Carbon (TOC). Petroleum hydrocarbons common to E&P wastes can be analyzed as total organic carbon (TOC) using wet oxidation technique. TOC is recommended as a quick and inexpensive screening technique for evaluating contamination levels for a large number of samples. Also, it can be used as a quality assurance/quality control (QA/QC) measure for O&G, TPH or GC-FID analyses. TOC frequency distribution for 1043 E&P reserve pits shows average, range, and distribution percentages compare favorably to previously discussed O&G and TPH test results (Table 3.18).

TABLE 3.16
FREQUENCY DISTRIBUTION FOR OIL AND GREASE
IN DRILLING RESERVE PIT SOLIDS

Concentration Interval (%)	Frequency Distribution	Percent of Total	Cumulative Percent
0–0.5	461	29.1	29.1
0.6–1.0	114	7.2	36.3
1.1–5.0	543	34.3	70.6
5.1–10.0	208	13.1	83.7
10.1–20.0	157	9.9	93.6
>20.0	101	6.4	100.0
Total	1584		
Maximum	99.1		
Minimum	<0.1		
Average	4.3		

TABLE 3.17
FREQUENCY DISTRIBUTION FOR TPH-IR
IN DRILLING RESERVE PIT SOLIDS

Concentration Interval (ppm)	Frequency Distribution	Percent of Total	Cumulative Percent
0–10,000	58	41.7	41.7
10,001–25,000	28	20.1	61.8
25,001–50,000	5	3.6	65.4
50,001–75,000	8	5.8	71.2
75,001–100,000	5	3.6	74.8
>100,000	35	25.2	100.0
Total	139		
Maximum	519,336		
Minimum	13		
Average	62,263		

3.2.8 Benzene

Total benzene frequency distribution results from analyzes of E&P wastes (including production sludges) (Table 3.19) show an average benzene level of 5.7 mg/kg for 134 separate E&P waste solids, ranging from <0.5 to 122.9 mg/kg. Almost sixty% of the waste solids analyzed would pass toxicity characteristic leaching procedure (TCLP) benzene threshold of 0.5 mg/kg by calculation (20 × 0.5), for example, 20 to 1 dilution.

3.3 PRODUCING OPERATIONS

Producing operations typically involve separating oil, water, and gas at a tank battery or satellite station. The separation process occurs in a closed system consisting of heaters, separators, free water knock-outs, and/or heater treaters. Generally, the only contamination occasioned from these pieces of equipment is due to leaks or purposeful draining of liquids to the ground.

Occasionally, a producing operation includes an emergency pit, and when used, contains fluids (oil and produced water, primarily) discharged from malfunctioning relief valves. These liquids are typically removed immediately after discharge to the pit.

The crude oil is stored in atmospheric tanks, equipped with equalizing lines. Nevertheless, over-topping the tank does occur. This results in oil saturating the soil. Similarly, produced water may be

TABLE 3.18
FREQUENCY DISTRIBUTION FOR TOTAL ORGANIC
CARBON (TOC) IN DRILLING RESERVE PIT SOLIDS

Concentration Interval (%)	Frequency Distribution	Percent of Total	Cumulative Percent
0–1.0	209	20.0	20.0
1.1–2.0	141	13.5	33.5
2.1–5.0	309	29.6	63.1
5.1–10.0	248	23.7	86.8
10.1–20.0	109	10.5	97.3
>20.0	27	2.7	100.0
Total	1043		
Maximum	58		
Minimum	<0.1		
Average	5.2		

TABLE 3.19
FREQUENCY DISTRIBUTION FOR TOTAL BENZENE
IN DRILLING RESERVE PIT SOLIDS

Concentration Interval (ppm)	Frequency Distribution	Percent of Total	Cumulative Percent
0–0.5	461	59.7	59.7
0.6– 5.0	114	15.7	75.4
5.1–10.0	543	6.7	82.1
10.1–20.0	208	9.7	91.8
20.1–50.0	157	6.0	97.8
>50.0	101	2.2	100.0
Total	134		
Maximum	122.9		
Minimum	<0.5		
Average	5.7		

spilled to the ground. Contaminants of concern include pH, sodium and hydrocarbons. The same manifestations are experienced in producing operations as with drilling operations. Refer to Drilling Operations this chapter for discussion of pH, sodium and hydrocarbon concentrations.

Produced water pits have not been studied as thoroughly as drilling pits, primarily because operators do not differentiate between pits during remediation. However, one operator identified a group of 51 production pits. The distribution of oil and grease in the pits solid phases ranged from < 1.0% to >12% (10,000 mg/kg to 120,000 mg/kg) (Table 3.20).

Typically, the operator skims the free hydrocarbons from the pit liquids. The pit liquids normally are injected into a Class II well. The produced water pH ranges from 3.4 to 8.7 with an average of 6.8 (Table 3.21).

3.4   IMPACT OF CONTAMINANTS

Heavy metals and sodium adversely impact humans and vegetation, either directly or indirectly. Studies show soluble chloride salts and excess exchangeable sodium are the major ions migrating in soil. These ions cause harmful effects on the soil and plant growth (Miller and Honarvar, 1975; Moseley, 1983). High soluble salt levels increase the osmotic potential in the soil lowering the amount of water available to plants from the soil. Also, increased osmotic potential interferes with the plant

TABLE 3.20
FREQUENCY DISTRIBUTION FOR OIL & GREASE
IN PRODUCTION PIT SOLIDS

Concentration Interval (%)	Frequency Distribution	Percent of Total	Cumulative Percent
0–1.0	17	33.3	59.7
1.1–3.0	8	15.7	75.4
3.1– 6.0	7	13.7	82.1
6.1–9.0	10	19.7	91.8
9.1–12.0	2	3.9	97.8
>12.0	7	13.7	100.0
Total	51		

TABLE 3.21
pH SUMMARY STATISTICS FOR API
PRODUCED WATERS

Test Statistic	Produced Waters
Sample size	17
Average (s.u.)	6.8
Median (s.u.)	6.9
Mode (s.u.)	6.8
Geometric mean (s.u.)	6.7
Variance (s.u.)	1.2
Standard deviation (s.u.)	1.1
Standard error (s.u.)	0.3
Minimum (s.u.)	3.4
Maximum (s.u.)	8.7

uptake of required nutrients (Kramer, 1969; Miller et al., 1980; Murphey and Kehew, 1984). High exchangeable sodium levels cause loss of soil structure, resulting in decreases in water and air infiltration and root zone soil compaction (Miller et al., 1980; Moseley, 1983).

Soils having high organic matter content and moderate percentages of clay are less sensitive to high sodium and soluble salts concentrations than coarser textured soils or loamy, or clayey soils in a dry environment (Miller, 1975; Moseley, 1983; Deuel, 1991).

Heavy metals found in drilling fluids include chromium, cadmium and lead. There is concern these metals might become incorporated and accumulated in the food chain or contaminate local groundwater, if leaching occurs from reserve pits. The availability (mobility) of metals in soil is dependent upon pH, reduction/oxidation potential and total metals concentration. *Leachate pH is the controlling variable.* In general, solubility of metals is inversely related to pH, except for barium, whose solubility is directly related to pH (Freeman and Deuel, 1984).

Migration of metal ions away from a pit site is limited by metal ion attenuation in clay minerals contained in the soil. Also, metal ions tend to form insoluble complexes in soil, especially at high pH levels (Moseley, 1983; Murphy and Kehew, 1984). Researchers find little or no migration of metal ions from reserve pits, because of clay attenuation and complexing (Henderson, 1982; Whitmore, 1982). Attenuation and migration of metals are affected by soil type and proximity of ground water. Migration is much more extensive in porous soils than in clayey soils and where ground water is close to the pit bottom (Murphy and Kehew, 1984).

Deuel and Holliday (1997) conducted a laboratory study of geochemical forms of heavy metals in E&P wastes to evaluate metal potential bioavailability and mobility in the environment. They concluded; low total and/or low exchangeable metal concentrations in E&P wastes control the bioavailability and mobility of heavy metals in soil.

REFERENCES

Deuel, L.E., Jr. 1991. "Evaluation of Limiting Constituents Suggested for Land Disposal of Exploration and Production Wastes." American Petroleum Institute (API), 1220 L Street NW, Washington, DC. 2005.

Deuel, L. E., Jr., and G.H. Holliday. 1997. "Geochemical Partitioning of Metals in Spent Drilling Fluids." Presented at ASME Energy Week, Houston, TX, January 28–30.

Freeman, B.D., and L.E. Deuel, Jr. 1984. "Guidelines for Closing Drilling Waste Fluid Pits in Wetland and Upland Areas." Presented at the 7th Annual Energy Sources Technology Conference and Exhibition. Sponsored by The Petroleum Division, ASME, New Orleans, LA, February 12–16.

Henderson, G. 1982. "Analysis of Hydrologic and Environmental Effects of Drilling Mud Pits and Produced Water Impoundments." Dames & Moore. API Production Department, Dallas, TX.

Kramer, P.J. 1969. "Plant and Soil Water Relationships: A Modern Synthesis". McGraw-Hill, New York.

Miller, R.W., and S. Honarvar. 1975. "Effect of Drilling Fluid Component Mixtures on Plants and Soil." Environmental Aspects of Chemical Use in Well Drilling Operation, conference proceedings, May, Houston, TX. EPA 560/1-75-004, pp. 125–143.

Miller, R.W., S. Honarvar, and B. Hunsaker. 1980. "Effects of Drilling Fluids on Soils and Plants: I. Individual Fluid Components." *J. Environ. Qual.* pp. 9:547–552.

Moseley, H.R. 1983. "Summary and Analysis of API Onshore Drilling Mud and Produced Water Environmental Studies." American Petroleum Institute, Bulletin D 19, Washington, DC.

Murphy, E.C., and A.E. Kehew. 1984. "The Effect of Oil and Gas Well Drilling Fluids on Shallow Groundwater in Western North Dakota." Report No. 82. North Dakota Geological Survey, Fargo, ND.

Rule 29-B. 2000. Title 43, Part 19, Chapter 3, §§ 311, 313, and 315—Pollution Control, State of Louisiana, Department of Natural Resources, Office of Conservation, Baton Rouge, LA, Promulgated January 20, 1986; revised October 20, 1990 and December 2000.

Whitmore, J. 1982. "Water Base Drilling Mud Land Spreading and Use as a Site Reclamation and Revegetation Medium." Fosgren-Perkins Engineering. APT Production Department, Dallas, TX.

SAMPLING METHODS AND PROTOCOLS

4.1 INTRODUCTION

Waste sampling is a vitally necessary first step toward site assessment successful site remediation. Improperly sampled sites provide unacceptable data on which to base remediation procedures. The purpose of this chapter is to discuss sampling tools, locations and protocols that will provide needed and accurate evaluation of the site.

The chapter does not address health and safety issues. Normally, these issues would be included in the operator's health and safety plan. However, note there is a potential for both chemical and physical health hazards during pit sampling activities. The extent of these hazards will vary from site to site and are minimized by using mechanical processes where possible; personnel training in safe work practices and limiting exposure to chemicals that impact health.

Being aware and taking personal responsibility for your safety is the best safeguard against physical health hazards. Similarly, attention to personal hygiene is a necessary safeguard against chemical assaults one might normally expect during pit closures. Chemical splashes and other dermal contact need to be mitigated as soon after an insult as possible. The recommended personal hygiene program includes:

- Wearing protective clothing, when necessary
- Immediately washing contamination from skin with soap and water
- Repairing rips and tears to protective clothing
- Washing hands before eating or smoking
- Washing hands and face after each shift

4.2 SITE INVESTIGATION

4.2.1 Sampling and Testing

The first step in waste disposal and pit closure is to sample and analyze pit materials. First, skim free oil from the liquid surface and, if possible inject the oil into the inlet of the production stream. Sample the aqueous liquid and analyze for salt content (EC) to determine if the water is fresh enough to be used for waste solids remediation processes. Remove the liquid, if too salty and commingle it with production waters for disposal in an injection well.

Site investigation procedures require a plan of action to maximize the safety of sampling personnel, minimize sampling time and cost, reduce errors in sampling, and protecting the integrity of the samples. Proper sampling procedures include:

1. SELECT proper sampling device
2. SELECT proper sample containers and closures

3. DESIGN sampling plan to provide:

   a. SAMPLING POINTS
   b. SAMPLE QUANTITIES, and
   c. SAMPLE COMPOSITS, if desired

4. IDENTIFY and PROTECT samples from spilling or tampering
5. RECORD sample information in field note book
6. Complete CHAIN of CUSTODY record
7. Complete LABORATORY SAMPLE ANALYSIS request sheet
8. DELIVER/SHIP samples to laboratory for analysis

4.2.2   General Guidelines

Before collecting samples, decide for which components analyses are required and why. For example, many exploration and production (E&P) operations generate multiphasic matrices consisting of both liquid and solid phases. Also, liquids may be multiphasic with a free-floating oil layer over water. Composite samples should reflect all phases in proportion to their occurrence. Phases to be treated, discarded and/or disposed of separately should also be analyzed separately.

Samples must be clearly labeled with the proper pit identification, type of sample, date, time and person taking the sample. It is important to distinguish between pit solids and the underlying native soils. A schematic showing sample location, depth interval and other pertinent information provides useful information when designing a pit closure plan. This information typically is retained field notebook.

4.2.3   Sample Containers

Collect samples in glass containers fitted with plastic or metal screw lids. One-quart or one-pint canning jars work well for all *inorganic* parameters. For organic parameters, such as, benzene or TPH, glass jars with Teflon®-lined lids are recommended. Each jar should be wrapped tightly with cellular foam materials to reduce breakage from jars colliding during shipment. Also, each jar should be placed in a sealable storage bag to avoid sample loss in event of jar breakage. Alternatively, soils or pit solids can be collected into sealable storage bags. However, extreme care must be exercised to avoid contamination of the locking device, which could cause sample leakage.

TABLE 4.1
WATER SAMPLE HOLDING TIMES, REQUIRED DIGESTION VOLUMES AND RECOMMENDED COLLECTION VOLUMES FOR METAL DETERMINATIONS

Measurement	Digestion Volume Required (ml)[a]	Collection Volume (ml)[b]	Preservative	Holding Time
Total Recoverable	100	600	HNO_3 to pH < 2	6 mo
Dissolved	100	600	Filter on site, NHO_3 to pH < 2	6 mo
Suspended	100	600	Filter on site	6 mo
Total	100	600	HNO_3 to pH < 2	6 mo
Chromium VI	100	400	Cool, 4~C	24 hr
Total mercury	100	400	HNO_3 to pH < 2	28 days
Dissolved mercury	100	400	Filter on site, NHO_3 to pH < 2	28 days

[a] Solid samples should be at least 200 g and usually require no preservation other than storing at 4° [1] C until analyzed. A 1-2 g subsample is digested and brought to the volume listed.
[b] Either plastic or glass. containers may be used.

Samples collected for "permitted discharge" or analysis by National Pollutant Discharge Elimination System (NPDES) permit criteria or USEPA hazardous waste characterization should be sampled in containers preserved and held per Table 4.1.

4.2.4 Sample Labeling

Label all samples immediately after placing the sample into the container. Use of a preprinted self-adhesive label facilitates labeling. DO NOT write directly on the glass; permanent ink will smear if the glass gets wet. Each label should contain a complete sample description, including date, time, sample location, sample description, and name of sampling personnel. *Confusion is avoided if the label includes the analytical parameters to be performed by the laboratory.* A chain of custody form or laboratory log-in sheet should be used to track the sample during sampling, delivery and analysis. *A chain of custody is preferred, if there is even a slight chance of the samples being involved in litigation.* Any sample used during legal procedures must have an accompanying chain of custody sheet signed by the sampler, transporter, receiver and laboratory personnel.

4.2.5 Sample Preservation

Sample storage and holding times need to be strictly followed. *Samples collected for Louisiana State-wide Rule 29-B analyses must be stored on ice while in transit to the laboratory.* Samples collected for "permitted discharges" or analyses by NPDES criteria or United States Environmental Protection Agency (USEPA) hazardous waste characterization must be held in accordance with EPA times, Table 4.1. Holding samples beyond the recommended time subjects the analytical results to question.

4.3   EQUIPMENT DECONTAMINATION

All field-sampling equipment should be laboratory cleaned, wrapped, and dedicated to a particular sampling site. If this is not possible, sampling equipment may be cleaned in the field, using the following cleaning strategy:

1. Nonphosphate detergent/tap water wash
2. Tap water rinse
3. Distilled/deionized water rinse
4. 10% nitric acid rinse[a] (trace metal or higher grade nitric acid diluted with distilled/deionized water)
5. Distilled/deionized water rinse[a]
6. Acetone (pesticide grade)[b]
7. Total air dry[b]
8. Distilled/deionized water rinse[b]

   [a] for metals/inorganic analysis
   [b] for organic analysis and TPH

Superscripts (a) and (b) refer to final rinse pathway. Distilled/deionized water (a) is used where mineral solutes and heavy metals are being analyzed. Pesticide grade acetone (b) follow distilled/deionized water rinse and equipment is allowed to air dry to remove traces of acetone.

4.4   SAMPLE COLLECTION

4.4.1 Sampling Water

Before collecting samples from a valve, spigot, or flow line, the sampling point must be flushed for a minute or two. Fill the sample container almost full. DO NOT OVERFILL the container. The exception to this rule is when sampling for volatile organics, e.g. benzene, toluene, ethylbenzene, xylenes, etc., which, require zero headspace for preservation of samples. In this case, overfill containers prior to sealing collects samples. *If zero head-space vials are used, after capping, turn the vial upside down. If gas bubbles are observed, up-end the vial and refill. Repeat the process until no bubbles are observed.*

To sample water from an auger hole or a well, a bailer is used. Insert the bailer into the hole until it reaches water. Allow the bailer to fill with water and collect the samples.

4.5   SOLIDS/SLUDGE SAMPLING PROTOCOL

Samples must be representative of the waste or impacted soil. In the case of pits, divide the pit into four to six sections, (approximately 30 m^2 each) depending on the pit size. This accomplished by placing stakes appropriately along edges and connecting the stakes orthogonally using twine, forming rectangles. Sample, within each rectangle, in an S-pattern or random pattern taking about ten samples per rectangle. Composite and retain the composite from each rectangular area. Form a pit composite, which is analyzed. If the analytical demonstrate a component of concern exceeds the Guidelines for Limiting Constituents (GLC), analyze the individual composite sample to establish from where the problem emanates. A similar technique can be use for impacted soil. I many case this technique allows special treatment of only the problem area. Refer to the below discussion regarding compositing.

4.5.1   Sampling Solids

Many different devices can be used to take waste and soil solids sampling. There are different types of augers, which are used depending on the waste/soil type. For sandy solids, a sand auger should be used (Figure 4.1). The teeth on the cutting end of the sand auger intersect so the loose solids don't fall out when the sample is pulled from the hole. For clays or drilling muds, a mud auger with open-sides works best (Figure 4.1).

A soil probe works well when the sample depth is less than 3 feet. Soil probes do not work well in sandy or compact soils. The sample size is very small with a soil probe (typically 3/4 in. in diameter); therefore, many samples must be taken and then composited.

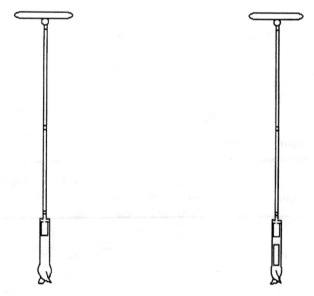

FIG. 4.1
A SAND AUGER IS DEPICTED ON THE LEFT AND A MUD AUGER ON THE RIGHT

Taking the sample   **Removing the sample**

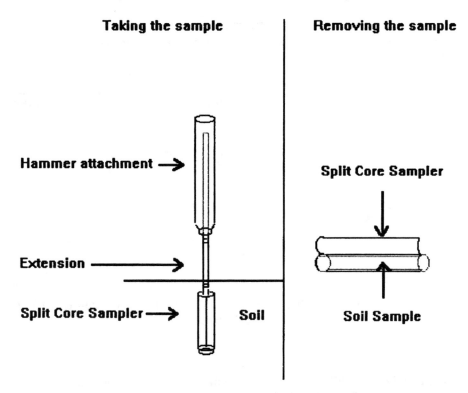

FIG. 4.2
TAKING A SAMPLE USING A SPLIT CORE SAMPLE

To take undisturbed a waste/soil core, a split core sampler is needed (Figure 4.2). Often a hammer attachment is needed to drive the split core into the ground. Disposable liners may be used and sent to the laboratory with the sample still inside the liner.

4.5.1.1 Sampling Pit Sludges. Sampling of sludges or sediments from storage pits, impoundments, ponds, or ditches usually is achieved by pushing a hollow 2-in. internal diameter tube [metal or polyvinyl chloride (PVC) plastic] into the pit bottom through all layers of waste, so the sample is representative of the waste matrix and impacted soil or sediment (Figure 4.3). Unconsolidated samples are retrieved by pushing the tube into consolidated native soil bottom to form a plug holding the sample in the tube. *Do not include the native soil plug as part of the sample.* An end cap or plugging device, such as a rubber stopper, valve or your hand can be placed on the top end of the tube after sampling. These plugging devices create a suction to keep the sample from discharging during tube retrieval. Viscous solids must to be discharged to the sample container with the aid of a smaller diameter ramrod. In large pit, a flat-bottom boat is needed to obtain access to all sampling points. Surface or

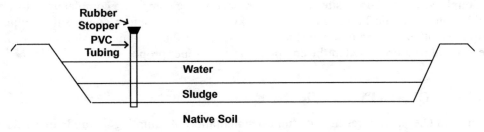

FIG. 4.3
REPRESENTATIVE SAMPLING OF PIT WASTES USING PVC TUBING

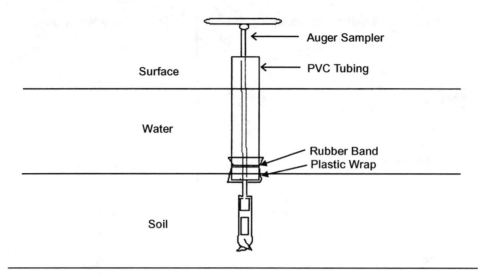

FIG. 4.4
**SAMPLING SOLIDS LOCATED BENEATH WATER WITHOUT CONTAMINATING
THE SOLIDS**

shallow samples can be taken with a scoop or shovel provided all layers are included in correct proportion to the sample matrix. Typically, all sludge samples and/or layers within cores are composited
in the field, see discussion below.

4.5.1.2   Sampling Solids Covered With Water.   Occasionally, it is necessary to sample solids located beneath water, such as a stream, river or lake bottom. Care must be taken to sample the solid
matrix only, without sampling the water. One method is to take a 6-in. diameter PVC tube, sealed
at the lower end with plastic wrap and rubber band, Figure 4.4. At the sample location beneath the
water, insert the tubing into the solids approximately 2 in. deep. Insert an auger into the PVC tubing
and puncture the plastic wrap. Use the auger to take the solid samples needed; being careful to keep
water from entering the tubing.

4.6   COMPOSITING SAMPLES

Samples should be composited on a weight/weight or volume/volume basis under controlled conditions. *Compositing should be limited when analyzing for volatile organic compounds, because compositing
causes a partial loss of volatile constituents.* After careful and complete homogenization of each individual solid sample, weigh equal portions of each individual sample into one container to form a composite sample. After the composite is made, homogenize the composite completely before analyzing.

Liquid samples should be mixed well by shaking or stirring before compositing. Equal portions
(measured as either volume or weight) are combined into one container. All phases of multi-phase
samples should be represented in the composite in the proper proportions.

4.7   FIELD QUALITY ASSURANCE

Proper field QA procedures include full documentation of sampling methods in a field notebook,
use of trip or travel blanks, and generation of blind split samples in the field. Refer to below discussion
for advantages derived trip blanks and spilt samples.

4.8   FIELD NOTEBOOK

Full documentation using a field notebook is required for all samples. Information should include date and time samples were taken, surface location, sample depth, sample matrix, soil type, and, if present, evidence of contamination (odor, color) or lack of contamination. Care must be taken to clearly identify sample containers.

4.9   TRIP BLANK

As a check for improperly cleaned sampling containers or trip contamination, each type of container used in the field on a given day should be filled in the laboratory with distilled water, sealed, and transported to and from the field with other sampling containers. The trip blank should then be analyzed alongside field samples.

4.10   SPLIT SAMPLES

Approximately 10% of the total samples taken for analyses (typically after homogenizing and compositing) should be split and placed into separate containers for separate analyses as a check on the laboratory procedures. These samples should not be identified as split samples. Although with soil and sludge samples, some variability can be expected between split samples, reasonable correlation is expected between samples.

In many cases, split sample are desirable if more than one party is involved in the sampling and analyses. In that case all of the individual samples are spilt and identified.

4.11   ELECTROMAGNETIC INDUCTION (EM) SURVERYS FOR LARGE VOLUME IMPACTS

Remote sensing techniques are available to evaluate soil quality following the spill or release of oilfield brine to agricultural land (Hokanson, 1995). Geonics manufactures two electromagnetic instruments, the EM-38 and EM-31 that have proven to be extremely useful in assessing soil salinity over relatively large volumes of field soil. The devices are used to survey land over a specified grid of appropriate density necessary to provide accurate contour maps of the area. The EM-38 and EM-31 devices are configured to define salt distributions from the surface to depths of 5 and 17 feet, respectively when operated in the vertical mode. EM-38 readings taken in the horizontal mode reflect salt distributions from the surface to 2.5 feet. The EM-31 device read in the horizontal mode define salt distributions from the surface to 10 feet.

4.12   SITE INVESTIGATION

4.12.1   Grid Layout

The grid is laid-out using pin flags or wooden stacks to define points to be assessed inclusive of background conditions and the area(s) of interest. Readings are made at each grid location and provide the basis for EC_x isolines used to map soil salinity of the area. Instrumentation equipped with data logger and GPS interface do not require a grid layout and are effectively mapped by taking continuous readings along transects set at a designed interval.

4.12.2   Soil Sampling and Analysis

Soil samples are taken at select grid locations in increments of 1 foot to a depth of 5 feet. Salinity is determined for each sample from electrical conductivity of the saturated paste extract as discussed in Chapter 6 and Appendix I (procedures are given in Appendix I, Anexes B and H). Surface samples are analyzed for sodium adsorption ratio (SAR), exchangeable sodium percentage (ESP) and cation exchange capacity (CEC) also discussed in Chapter 6 and Appendix I (procedures given in Appendix I, Anexes G, F and D). SAR, ESP and CEC are used to determine treatment strategies and amendment levels necessary to abate salt and sodicity as discussed in Chapter 8.

4.12.3   Statistical Analyses

A significant aspect of the EM-38 device is that it provides a weighted average of soil salinity similar to the way plants draw moisture from the soil profile and thus respond to salts in transpired water. Research as shown that the 0-1 ft, 1-2 ft, 2-3 ft, 3-4 ft and 4-5 ft depth intervals contribute about 43, 21, 10, 6 and 10% respectively to the conductivity reading of the EM-38 device (Rhoades and Corwin, 1980). These percentages serve as coefficients to calculate the profile EC used to ground truth EM-38 readings. The computing formula is as follows:

$$\text{Profile EC} = 0.43*\text{EC } 0\text{-}1' + 0.21*\text{EC } 1\text{-}2' + 0.1*\text{EC } 2\text{-}3' + 0.06*\text{EC } 3\text{-}4' + 0.1*\text{EC } 4\text{-}5'$$

where EC is measured for a saturated paste extract from the depth interval indicated. Hokanson (1995) describes in detail the 4 steps required to statistically validate EM survey results:

- Calculate profile EC values from soil core analyses
- Generate regression models correlating profile EC and EM readings at test locations
- Calculate predicted profile EC values using various regression models
- Accept the regression model that more accurately predicts measured profile ECs

4.12.4   Results

The resulting contour plot provides a detailed map of salinity with the survey area. The survey also serves as a guide to place soil borings necessary to ground truth remote sensing instrumentation and define soil conditions and parameters used to assess soil quality. The resulting map often serves as a field guide in remedial actions plans.

Interferences. The reader is alerted to interference and uncontrolled variability in readings caused by buried metal. An experienced operator can generally detect such interference and work around it. However, the authors have experienced areas where the interferences are too great to manage restricting the utility of the instrumentation.

REFERENCES

Hokanson, S. 1995. Using electromagnetic induction surveys to delineate salt impacted soils: Statistical correlation to soil core data. 2nd Annual International Petroleum Environmental Conference Proceedings. New Orleans, Louisiana, pp. 651–665.

Rhoades, J.D. and D.L. Corwin. 1980. Determining soil electrical conductivity-depth relations using an inductive electromagnetic soil conductivity meter. Soil Sci. Soc. Am. J. 45:255–260.

Rule 29-B, 2000. Title 43, Part 19, Chapter 3, §§ 311, 313 and 315—Pollution Control, State of Louisiana, Department of Natural Resources, Office of Conservation, Baton Rouge, LA, Promulgated January 20, 1986; revised October 20, 1990 and December 2000.

SAMPLING PROGRAM

5.1 LARGE FACILITY IMPOUNDMENTS

Think of small pits as one or more areas of a larger pit and sample accordingly. Large facility impoundments, typically, receive the full spectrum of E&P wastes, which can be broadly classified into three major groups:

1. Weathered drilling fluids and drill cuttings from past operations
2. Recent spent drilling wastes reflecting newer formulations and/or
3. Tank bottoms, and other produced liquids and sludges

Impoundments source variability, for example, size, and natural segregation of materials with depth and distance from the point of discharge, presents a challenge for obtaining representative samples. Representative samples are fundamental for characterization of pit content parameter variability. This variability, in turn, is necessary in establishing the scientific validity and relevance of the analytical measurements.

5.2 STATISTICAL VALIDATION

We selected a large commercial waste pit to verify waste distribution within pits. The pit was divided into 15 areas (Figure 5.1). Each area was subdivided into four equal sampling sites, Figure 5.1. Using a flat-bottom boat and sampling tube, random duplicate sample cores were taken to native substrate or 3 meters, whichever came first. Each sample was placed into a glass container and labeled as to area, sampling site and replicate. Each sample was homogenized and analyzed for pH.

An analysis of variance (ANOVA) of pH showed pit material pH varied significantly between areas and between sampling sites (Table 5.1). The F-test shows a significant interaction between areas and sampling sites (Table 5.1). The analysis demonstrates no significant interaction between replicate samples.

The analysis shows pH values, averaged over each replicate sample area, are much higher values at the off-loading ramp (Table 5.2). Also, Table 5.2 shows the results of a Tukey multiple range test comparing mean pH values between sampling sites and zones. Results show unit 1 yielding the highest pH (Table 5.2), followed by area 2, which was significantly greater than areas 3 and 5. Area 4 was significantly lower in pH than the aforementioned areas, but was statistically equivalent to all others. Site means averaged over replicates and areas reveals sites 1 and 2 significantly higher in pH than sites 3 and 4.

A three-dimensional surface plot shows the interaction between area and site and the effect of this relationship on pH (Figure 5.2). The fact inner units are not significantly different suggests that the large E&P facilities impoundments could be sampled close to the sides. One should sample a sufficient

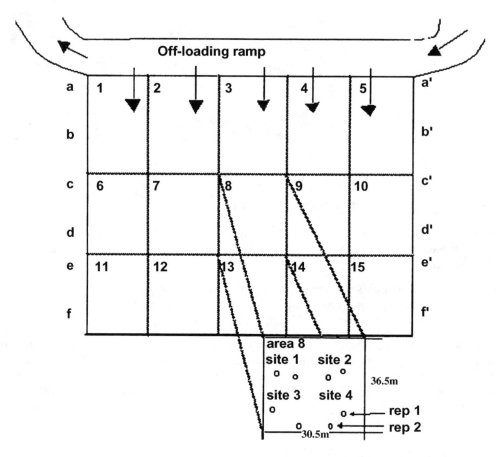

FIG. 5.1
SCHEMATIC SHOWING PIT DELINEATED BY AREA, SITES WITHIN AREAS, AND REPLICATE CORE LOCATIONS

distance from the bank such that the depth of the materials encountered and the potential for layering is representative of the pit as a whole.

Figure 5.2 graphically displays pH highest near the offloading ramp (areas 1–5) and decreases with distance away from the ramp. Apparently, the trucks pull to the northwest corner of the pit during

TABLE 5.1
THE ANOVA FOR pH IN LARGE PIT SAMPLES

Source	df	SS	Mean Square	F-ratio
Treatments				
Reps	1	0.0078	0.0078	0.067
Unit (u)	14	76.5699	5.4693	47.009[a]
Site (s)	3	13.9694	4.4693	43.023[a]
u x s	42	39.6021	0.9429	8.104[a]
Error	59	6.8644	0.1163	
Total	119	137.0138		

[a] Denotes highly significant treatment effect, p < 0.01.

TABLE 5.2
MEAN pH VALUES BY UNIT AND SITES

	Site				Site[a]
Area	1	2	3	4	Average
1	11.17	10.73	10.55	8.39	10.21a
2	11.00	11.55	7.68	7.64	9.47b
3	9.71	9.52	7.87	7.51	8.65c
4	7.81	7.82	7.53	7.48	7.66d
5	8.12	10.03	7.57	8.09	8.45c
6	8.53	7.68	7.29	7.27	7.69d
7	7.81	7.71	7.97	7.62	7.78d
8	7.49	7.64	7.45	7.49	7.52d
9	7.52	7.23	7.55	7.43	7.43d
10	7.40	7.64	7.50	7.39	7.48d
11	7.78	7.85	7.62	7.84	7.76d
12	7.98	7.79	7.52	7.46	7.69d
13	7.52	7.75	7.44	7.56	7.57d
14	7.66	7.59	7.46	7.55	7.56d
15	7.39	7.43	7.59	7.46	7.47d
Average*	8.32a	8.40a	7.77b	7.61b	

[a] Averages followed by the same letter are not significantly different at $P \leq 0.05$.

E&P Facilities Pit

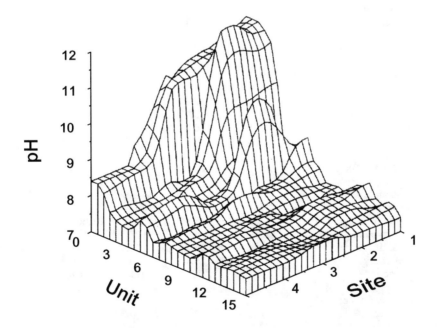

FIGURE 5.2
SURFACE PLOT DEPICTING THE AREA SITE INTERACTION

unloading since the pH is highest in areas 1 and 2. Further, when several trucks unload, the second truck appears to unload near the northeast corner (area 5, site 2) of the pit. We conclude the distance from the point of discharge has a significant effect on waste composition.

5.3 ZONAL COMPOSITES

Transect composite samples were prepared by combining subsamples of each core taken along transects. Zonal composites 1, 2 and 3 were prepared by combining subsamples of transect composites a–a', b–b'; c–c', d-d'; e–e' and f–f', respectively, Figure 5.1. Zone 1 composite is equivalent to combining all replicate subsamples from areas 1–5. Zone 2 represents areas 6–10 and zone 3 represents units 11–15.

Zonal composites were analyzed for the suite of the significant parameters, Table 5.3, to evaluate materials composition and distribution relative to the point of discharge. The moisture content of solids was highest in Zone 3 materials, demonstrating the finer dispersed particles tend to be segregated with increased distance from the point of discharge. This was reflected in the higher cation exchange

TABLE 5.3
ZONAL COMPOSITE ANALYSES

Parameter[a]	Zone 1	Zone 2	Zone 3
Moisture, %	73.7	152.6	211.6
pH, s.u.	9.7	8.6	9.7
SPEC, mmhos/cm	11.1	9.4	10.1
1:1 Soluble cations, mEq/l			
Sodium	62.5	59.9	60.5
Calcium	11.4	4.0	9.9
Magnesium	<0.1	0.5	<0.1
Potassium	<0.1	<0.1	<0.1
SAR, unitless	26.1	39.9	27.1
Exchangeable Cations, mEq/100g			
Sodium	1.1	4.9	8.3
Calcium	2.2	4.8	4.8
Magnesium	0.6	2.2	2.4
Potassium	<0.1	<0.1	<0.1
ESP,%	4.8	15.6	23.3
CEC, mEq/100g	22.8	31.1	35.7
Total metals, mg/kg			
Barium	2326	1307	1051
True total barium	179,816	103,030	87,864
Chromium	74	127	190
Lead	62.2	59.1	61.8
Mercury	0.1	0.1	0.3
Zinc	241	211	207
Organic fraction			
TPH, %	5.6	4.2	5.3
TOC, %	3.8	4.3	3.7

[a] Analysis methods are from Laboratory Procedures for Analysis of OilfieldWaste, Louisiana DNR, August, 1988.

capacity (CEC) and greater exchangeable sodium percentage (ESP) of the Zone 3 solids. Soluble phase constituents including electrical conductivity (EC), and cationic speciation were approximately equal in all three zones. Also, total petroleum hydrocarbon (TPH) was approximately equal in all three zones. Total metal compositions varied considerably between zones and apparently between sources. Barium, for example, was highest in Zone 1, while chromium was highest in Zone 3. This suggests that barium is related more to the coarser fraction of the waste as an additive, and chromium exists more as a tramp contaminant of clay.

5.4   LARGE-PIT SAMPLING CONSIDERATIONS

Based on the materials composition and distribution of solids in the large impoundments, it is clear that a minimum of two samples should be generated for each pit. One approach to collecting the two samples is given in Figure 5.3, for example, divide the pit into two zones paralleling the ramp. An aluminum boat or even a crane with a basket platform can be used to collect cores. The main objective is to sample far enough into the pit to obtain a vertical profile representative of the waste solids stratification.

Duplicate cores should be taken from five to seven sampling points within each zone and blended together to form Zone 1 and Zone 2 composites. Pit characterization should include analysis of both composites.

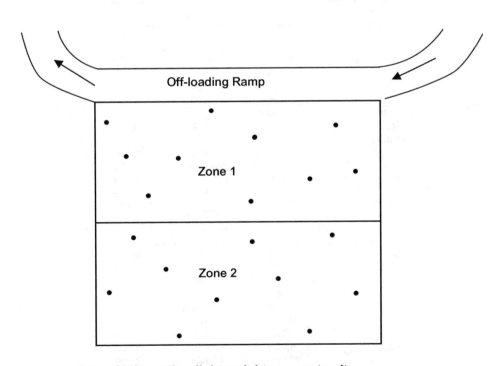

• Boat, Skif, or other light weight access to pit.

FIGURE 5.3
SCHEMATIC SHOWING RECOMMENDED SAMPLING POINTS

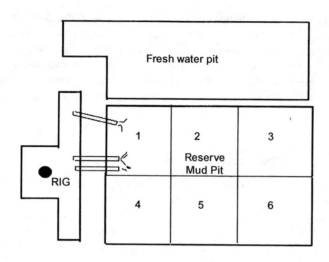

Section	Dry Solids (bbls)
1	4,491
2	3,505
3	2,485
4	2,060
5	2,292
6	1,835

FIGURE 5.4
SCHEMATIC OF RESERVE PIT IN OPERATION

TABLE 5.4
METAL PARAMETER ANALYSES BY SECTION

Metal[a] Parameter	Section		
	1	2	3
Barium	38,507	39,761	12,000
Chromium	34	20	15
Lead	60	56	35
Zinc	184	230	87
Arsenic	6.6	5.0	2.9
Iron	21,560	15,976	10,782
Manganese	391	202	191
Sodium	1,600	2,800	2,680
Calcium	38,600	20,000	9,600
Magnesium	4,290	3,040	2,500
Potassium	3,923	3,510	3,132

	Section		
	4	5	6
Barium	26,992	6,317	6,143
Chromium	26	24	29
Lead	52	32	37
Zinc	32	12.4	13
	6.4	4.9	4.8
	20,946	17,725	16,555
	221	213	209
Arsenic	4,920	5,660	5,030
Iron	22,800	15,100	12,200
Manganese	4,310	4,110	4,140
Potassium	4,220	4,135	4,117

[a] Analysis methods are from Laboratory Procedures for Analysis of Oilfield Waste, Louisiana DNR, August 1998.

5.5   RESERVE PITS

Reserve pits are identified as earthen impoundments used to receive E&P wastes during the drilling process. They are much smaller than the facility impoundments and therefore more easily sampled in representative fashion.

5.6   SAMPLING STRATEGY

A reserve pit receives and holds wastes from shakers, hydrocyclones and centrifuges (Figure 5.4). For sampling, the pit is subdivided into six Sections (Figure 5.4). The sections to be sampled are accessed by a flat-bottom boat. Solids and liquids samples are collected from 10 sites within each section by pushing a hollow tube through pit contents and into consolidated native soil defining the pit bottom (Figure 4.3). The liquids are discarded, because liquids typically contain less significant parameters than solids. Site solids cores are combined to form a section composite. Section composites are sub-sampled individually into glass containers prior to be blending into a pit composite sample.

5.7   CORE COMPOSITE VALIDATION

Several section samples were analyzed for metal parameters to determine the metals distribution within the reserve pit near the point of waste discharge (Table 5.4). All metals, *except sodium*, were highest in concentration within the sections nearest the points of discharge (Figure 5.4), for examples, sections 1, 2 and 4. Sodium was distributed in reverse to the trend of the other metal. Sodium concentrations were highest in sections 5 and 6. Sodium reflects the dispersed nature of the clay gels used in drilling fluids, with the fine particulates tending to remain in suspension over a longer distance from the point of discharge. Calcium saturated clays tend to be flocculated and fall out with coarser clay particulates and other solid additives (Table 5.4).

Weighted averages calculated from individual section sample analyses are in close agreement with the overall pit composite (Table 5.5). The data demonstrate multiple core composites provide an acceptable means of obtaining representative samples from both aerial and spatially variable matrices. Composites also provide a means for reducing analytical effort and associated costs.

TABLE 5.5
COMPARATIVE PIT SOLIDS ANALYSIS

Metal[a] Parameter	Section Average mg/kg	Pit Composite mg/kg
Barium	25,406	29,191
Chromium	25.3	16
Lead	47.0	64
Zinc	118	120
Arsenic	5.3	3.1
Iron	17,624	15,100
Manganese	256	289
Sodium	3,360	3,450
Calcium	22,275	26,980
Magnesium	3,722	4,110
Potassium	3,375	2,995

[a] Analysis methods are from Laboratory Procedures for Analysis of Oilfield Waste, Louisiana DNR, August 1988.

5.8   FUTURE SAMPLING CONSIDERATIONS

In general, reserve pits should be subdivided into approximately 30 m^2 areas sections, with each section represented by 5–10 site composites. A 60 m^2 pit yields two section samples; a 90 m^2 pit generates three section samples; etc. Section composites are transported to the laboratory in separate containers so pit composites can be prepared at the laboratory in proportion to their occurrence. Typically, pit composites are analyzed for the full suite of parameters deemed necessary for waste characterization. Section samples are analyzed for only those constituents requiring management for proper disposal (see Chapter 7).

PROTOCOL FOR
CHEMICAL ANALYSES

6.1   SOLUBLE CONSTITUENT ANALYSES

Pure water is *not* a good conductor of electricity; however, water containing soluble salts will conduct electricity roughly in proportion to the quantity of salt present. This relationship is accurate enough to determine the salt concentration by measuring the electrical resistance or the electrical conductivity of a solution. In this procedure, the electrical conductivity of a soil-saturated paste extract is used to estimate the amount of soluble salts in a waste/soil.

To determine the soluble salts, the soil is extracted with distilled or deionized water. The salts dissolved will increase with increasing waste/soil:water ratio. Salt determinations normally are based on conductivity of an extract of a waste/soil when at saturation, called saturated paste extract.

Measurements of soluble salts by this method are accurate and reproducible in the range usually occurring in the saturation extracts. There is a high degree of positive correlation between saturated paste electrical conductivity (SPEC), total concentration of cations or anions, and the osmotic pressure of soil-water extracts. However, due to differences in equivalent weights and equivalent conductivities of the different ions, the relationship of conductivity to salt content on a weight-to-weight basis is poor. The following SPEC relationships are employed:

- Soil/waste salt concentration, mg/L = 613 × saturated paste electrical conductivity (SPEC), millimhos/cm (mmhos/cm). The constant 613 produces very good correlation for oil and gas field wastes
- Total cation concentration, meq/L = 10 × saturated paste electrical conductivity (SPEC), mmhos/cm
- Osmotic pressure, atmospheres (amt) = 0.36 × saturated paste electrical conductivity (SPEC), mmhos/cm

The standard unit for conductivity, mhos/cm is too small for field use. Accordingly, for convenience millimhos/cm is used. Units commonly reported are:

$$0.001 \text{ mhos/cm} = 1 \text{ millimhos/cm} = 1000 \text{ micromos/cm} = \text{msiemens/cm}$$

6.2   PREPARATION OF SATURATED PASTE AND EC ANALYSIS

Place 100 g of oven dried soil/waste into a 250-ml beaker. Add water to the soil in small increments from a graduated cylinder. Successive increments of water are added with slow stirring (*without vigorous stirring*) of the soil until water just wets the entire soil/waste mass. Then a few drips more of water are added slowly until the surface glistens slightly. After this moisture content has been reached, the soil/waste is stirred using a glass rod and drops of water are added until the soil is a "paste" which just barely flows together to close around a hole in the paste left by the rod. The soil/waste is now at the moisture saturation%age. The surface of the water saturated soil/waste glistens and the soil/waste air has been excluded. *Record the amount of water used to reach this point.* Calculate the% of moisture at saturation using:

$$\% \text{ Moisture at saturation} = (\text{wt soil water}/\text{wt oven dry soil}) \times 100$$

Allow the soil/waste to stand for about 30 minutes to equilibrate and transfer to a Buchner funnel fitted with filter paper. Apply suction and collect at least 10 ml of water extract. Take the temperature of the extract and set the proper correction on the conductivity meter. Transfer the extract to a 50-ml beaker. Rinse the meter cell with distilled water several times and with the soil-water extract once before taking a conductivity reading of the extract. Obtain a reading on the solu-bridge. Report the conductivity of the soil/waste extract in millimhos/cm (mmhos/cm). Further details on obtaining the saturated paste extract and EC analysis are given in Appendix I, Annexes H and B, respectively.

6.3   PERIPHERAL ANALYSES AND RELATIONSHIPS

The "equipotential" moisture status associated with a saturated paste condition represents the highest likely moisture level attainable in the field. It is a reproducible value, but requires some training to perfect the technique. Unskilled technicians tend to puddle the soil/waste by adding too much water. Puddled soils/wastes imbibe water to varying degrees causing considerable deviation in replicate parameter analysis. Drilling wastes received by the laboratory for analyses often contain smectitic clays in a puddled condition. *Puddled soils can have in excess of 200–300% moisture and represent a significant dilution of salts and grossly underestimate of the salinity hazard associated with the waste.* To avoid this problem, a subsample of drilling waste should be air-dried and then moistened to a saturated paste level or extracted at a 1:1 waste: water ratio (by weight) for soluble constituent analyses.

Some matrices may be hydrophobic (repel water) due to the presence of petroleum hydrocarbon, and require heat treatment for hydrocarbon removal prior to sample preparation for soluble constituent analyses. For details on testing for hydrophobicity and heat treatment see Appendix I, Annex E.

6.4   SOIL/WASTE REACTION

Soil/waste reaction is an estimate of the hydrogen ion concentration as measured with a combination glass electrode and pH meter. The procedure for determining pH is given in Appendix IV, Annex B. A saturated paste provides a thick moisture film and good contact with the glass electrode. The normal soil pH range for good growth of agronomic crops is 5.5 to 8.3. Generally, soils/waste having pH below 5.5 develop toxic aluminum and manganese levels and soils pH above 8.3 reflect a sodium bicarbonate/carbonate buffer system resulting in excessive and plant-toxic soluble sodium levels. Many native plants adapt to soils with a pH range of 4–9. *Nevertheless, we recommend working to a pH range of 5.5–8.3.* Excessive soil/waste reactions are readily treated by amendment (Chapter 8).

Excessive alkalinity (pH > 8.3) typically accompanies sodic soil/waste conditions. This condition requires sulfur amendment to lower the soil reaction, and gypsum to reduce sodium saturation in the adsorbed phase. Excessive acidity (pH < 5.5) is controlled by lime addition. Procedures for determining the alkalinity and acidity of drilling wastes are given in Appendix IV, Annexes C and D, respectively.

6.5   SOLUBLE CATIONS

Sodium (Na), potassium (K), calcium (Ca) and magnesium (Mg) are the cations of interest in soil/waste. They are measured using saturated paste extracts and direct aspiration atomic absorption spectroscopy. Sodium and potassium are analyzed in a flame emission mode. Calcium and magnesium are analyzed in the absorbency mode. Results are normally reported in milliequivalents per liter (meq/L).

Sodium adsorption ratio (SAR) is calculated from saturated paste concentrations of Na, Ca and Mg expressed in meq/L:

$$SAR = Na/[(Ca + Mg)/2]^{1/2}$$

SAR is used in conjunction with EC to evaluate the potential hazards associated with sodium and other soluble salts. A soil/waste with an SAR value >12 tends toward a sodic condition. *A sodic soil/waste is defined as one with ESP >15%, for example, exchange sites occupied by sodium, and SAR > 12.* The U.S. Salinity Laboratory staff (1954) developed an empirical equilibrium relationship relating SAR to ESP:

$$ESP = \frac{100\,[-0.0126 + 0.01475 \times SAR]}{[1 + (-0.0126 + 0.01475 \times SAR)]}$$

A sodic condition reflects an undesirable physical status (a structureless or dispersed soil/waste condition). The procedure for determining soluble cation distributions from which SAR is calculated is given in Appendix I, Annex G.

The degree of hazard to plants is related to the ionic strength of the matrix. Electrical conductivity (EC) > 4 mmhos/cm is classified as saline. *Soils/wastes with a high ESP [≥15% and low salinity level (EC <1)] are readily ameliorated with gypsum amendments. When both ESP and EC values are high (>15 and >4, respectively), the soil is considered saline-sodic and requires both gypsum and leaching to restore productivity.*

6.6 SOLUBLE ANIONS

Bicarbonates, carbonates, hydroxides, chlorides and sulfates are the anions of interest in soil, drilling waste and leachate extracts. When color changes are readily visible in the various water extracts, bicarbonate, carbonate, and hydroxide ions are titrated in the presence of phenolphthalein and methyl orange indicators. Water extracts from sodic soils and some drilling fluids are often dark colored and require titrations by a pH meter instead of by indicators. Titration of an alkaline solution to the phenolphthalein end point (pH 8.2) with a standard solution of strong acid provides a measure of carbonate, and further titration to the methyl orange end point (pH 4.5) provides a measure of bicarbonate. Chloride is measured for noncolorgenic soil/waste extracts and water samples with an argentometric titration using potassium chromate as the indicator. The potentiometric method is used for colored or turbid extracts. Drilling waste extracts containing chromates and organics may require pretreatment by "soft digestion" to remove interferences prior to titration. Sulfate is determined for soil and drilling fluid extracts using the turbidometric method. Make background correction for color or turbidity by subtracting the absorbency values in the absence of barium chloride from the absorbency in the presence of color.

6.7 QUANTITATIVE AND QUALITY CONTROL

Quality assurance and the confidence in the reliability of analytical results for soluble constituents begin with deionized water in standard formulations and dilutions, and as a final rinseate for all laboratory glassware. *Scrupulous cleaning of all glassware starts with a prerinse scrubbing with soap and water, rinsing with distilled water, dilute nitric acid, and finally deionized water.*

Analytical and electronic balances must be calibrated and checked against standard weights prior to and during use in gravimetric assays. Calibrated glassware is used in volumetric procedures. Precision of direct reading instruments is maintained by routine calibration against commercially available buffer solutions and standards.

Cations and anions are determined by indirect methods employing a comparison of the sample matrix response to a standard curve prepared by known addition technique. Precision and accuracy are evaluated statistically with replicate determinations and spike recovery analyses. Detection and sensitivity limits are evaluated with reagent blanks and standard deviation about the mean for the low standard. Matrix interferences for alkali metal determinations are controlled by the addition of suppressants to aid in element ionization and surfactants that maintain optimum atomization rates into the instrument flame.

6.8   SOLID PHASE ANALYSES

Colloidal clay and organic particles combine to form the chemically active components of waste solids and soils. Colloidal size fractions have very large surface areas. Additionally, clay separates have an electrical charge [cation exchange capacity (CEC)] emanating from ionic substitution and broken crystal edges. Soil/waste clays are comprised of crystalline sheets of silicon oxides and aluminum oxides or hydroxides bound to each other by shared oxygen atoms. As discussed in Chapter 2, a charge deficiency within the crystalline lattice arises when aluminum (Al^{+3}) replaces silicon (Si^{+4}) in a tetrahedral sheet, or iron (Fe^{+2}), zinc (Zn^{+2}) or magnesium (Mg^{+2}) replaces aluminum (Al^{+3}) in an octahedral sheet. The negative charge deficiency attracts and retains cations. Cations are free to exchange between soil/waste solution and the adsorbed phase. *The negative charge per unit soil mass is termed the cation exchange capacity (CEC)*. Equilibrium exists between soluble and adsorbed ions, since cation exchange is rapid and reversible. *The equilibrium relative to the solid phase is termed the exchangeable cation distribution.*

6.9   CATION EXCHANGE CAPACITY (CEC)

Soil/waste colloids have both negative and positive charged surfaces. Capacity to adsorb cations per unit mass is referred to as the cation exchange capacity (CEC), and is expressed in milliequivalents per 100 g (meq/100g) soil/waste mass on an oven dry (105°C) weight basis. *The capacity to adsorb anions is called the anion exchange capacity (AEC) and is expressed in milliequivalents per 100 g soil/waste mass.*

The two most common measures of CEC are the quantity of cations adsorbed from a salt solution buffered at either pH 7 or 8. CEC determinations from salt solutions buffered at pH 8 give an accurate assessment of the charge at the clay surface and a reasonable measure of contributions from organic matter and noncrystalline colloids. *Most drilling wastes are alkaline in reaction and therefore best evaluated using pH 8 buffered salts.*

6.10   EXCHANGEABLE CATION DISTRIBUTION

The exchangeable cations in a soil are the positively charged ions held on the surface exchange sites and in equilibrium with the soil/waste solution. *The major cations, Na, Ca, K and Mg, are called basic cations and the percentage of the CEC occupied by these cations is called the percentge base saturation.* Fertile soils approach 80% base saturation, with bulk saturation distributed mainly as Ca and Mg.

Problems arise when the exchangeable Na saturation percentageage [$100 \times (Na/CEC)$] exceeds 12–15% of the CEC. *These are termed sodic soils and are characteristically structureless and impervious to air and water.* Many drilling fluids (water base in particular) are dispersed with sodium hydroxide. *Adsorbed sodium management becomes a major concern in produced water and water base and salt-water drilling fluid wastes site reclamation.*

6.11   ANALYTICAL PROCEDURES

The procedures for determining CEC and exchangeable cation distributions are given in Appendix I, Annexes D and F, respectively.

6.12   QUALITY CONTROL AND ASSURANCE

Accuracy and precision of CEC and exchangeable cation distribution measurements are evaluated at a procedural level, since they are influenced by treatment parameters, including incorporating of waste streams into natural soil bodies. *Accuracy assessment of CEC results from multiple determinations of laboratory reference materials representative of the matrix in question, and previously evaluated by several techniques. Precision is achieved by replicate determinations.*

The three steps of CEC determination are:

1. Saturation of cation exchange sites with a specific cation;
2. Removal of excess saturating solution; and
3. Replacement of saturating cation.

Errors can arise at each of the three steps. Step 1 errors are minimized through selection of appropriate saturating cation and adequate rinsings with index cation solution. Many drilling wastes contain adsorbed sodium, rendering it the most suitable index cation. Step 2 errors are minimized by thoroughly washing of the sample with deionized water followed by isopropyl alcohol rinse. Using isopropyl alcohol in the final rinse process reduces loss of sediment on decantation experienced with a total deionized water washing. Errors can arise during Step 3 from relatively unweathered soils samples due to carbonates or gypsum when calcium or magnesium is used as index and replacing cations. A sodium index and nitrate-replacing agent negates possible error in step 3 caused by carbonates or gypsum.

6.13   TOTAL METALS ANALYSES

Most soils and drilling wastes contain low quantities of heavy metals in water soluble or exchangeable form. *Accordingly, water or other extractants that partition (separate from the matrix) these fractions do not provide a suitable index for metals management. On the other hand, total metal analysis provides usable numbers and a basis for comparison with acceptable total metal loading criteria.* Refer to **API Publication Number 4600** for technical guidance associated with land-managed drilling wastes.

6.13.1   Acid Digestion

The recommended protocol for total metal digestion is detailed in Appendix II. Metals fractionating procedure estimate all total metals concentrations, except barium, accurately. A separate procedure was developed for true total barium (Appendix II).

6.13.2   Quality Control and Assurance

Matrix variability is managed by careful blending prior to weighing subsamples for analysis. Certified reference soils and sediments are available along with duplicate digests and matrix spikes to evaluate precision and accuracy of analysis.

6.14   ORGANIC ANALYSES

The oil and grease (O&G) procedure (Appendix I, Annex A and TPH-IR (EPA 418.1 as modified for soils and waste solids) provide an index of petroleum hydrocarbon partitioned from the matrix by a solvent. The O&G method uses methylene chloride and the TPH-IR procedure uses freon. Sample matrices are acidified to reduce co-extraction of naturally occurring organic matter. Additionally, solvent extracts are often passed through silica gel to remove interferences prior to concentration and analysis. Freon is in short supply and may not be available at all in some states. As an alternative we recommend substituting a TPH analysis based on EPA SW-846 Methods 8000B, 8015C and 8100 over an extended carbon range C_6 to C_{35}.

Either of these analyses provides a reasonable estimate of total petroleum hydrocarbon. Values are used to determine hydrocarbon-loading rates with respect to on-site land treatment or bioremediation, and also serve as an index to other management inputs. Specific component analyses may be necessary for permit purposes, depending on drilling waste source and regulatory agency involved.

Neither O&G nor TPH-IR serves as an adequate index for tracking bioremediation treatment performance. Both methods give inflated petroleum hydrocarbon values. We believe the reason for this is the conversion of petroleum hydrocarbon substrates to co-extracted biomass. Gas chromatographic analysis of total petroleum hydrocarbon *standardized to peak area response* is a better indicator of bioremediation performance (Deuel and Holliday, 1993).

6.14.1   Quality Control and Assurance

Duplicate analysis, spike recovery analysis, reagent blanks and methods blanks are necessary to evaluate precision and accuracy of the analyses performed. Selection of standards of close similarity to the source hydrocarbon is critical to accurate hydrocarbon analyses. This is particularly true when comparing peak area response in gas chromatography.

REFERENCES

American Petroleum Institute. 1995. Metals Criteria for Land Management of Exploration and Production Wastes: Technical Support Document for API Recommended Guidance Values. API Publication No. 4600. Washington, DC.

Deuel, L.E., Jr., and G.H. Holliday. 1993. "Determining Total Petroleum Hydrocarbons in Soil." Presented at the 68th Annual Technical Conference and Exhibition of the Society of Petroleum Engineers SPE 26394, held in Houston, TX, October 3–6.

U.S. EPA. 1978. Total Recoverable Petroleum Hydrocarbon by Infrared Spectrophotometer. Method 418.1. EPA 600/78/001. USEPA, Washington, DC.

EPA, 1983. "Methods for Chemical Analysis of Water and Wastes," 3rd Edition, Environmental Protection Agency, Environmental Monitoring Systems Laboratory-Cincinnati (EMSL-Ci), Cincinnati, Ohio 45268, EPA-600/4-79-020, Method 418.1.

U.S. EPA. 1986. Polynuclear Aromatic Hydrocarbons. Method 8100. SW-846 Test Methods for Evaluating Solid Waste. USEPA, Washington, DC.

U.S. EPA. 1996. Determinative Chromatographic Separations. Method 8000B. SW-846 Test Methods for Evaluating Solid Waste. USEPA, Washington, DC.

U.S. EPA. 2007. Nonhalogenated Organics by Gas Chromatography. Method 8015C. SW-846 Test Methods for Evaluating Solid Waste. USEPA, Washington, DC.

U.S. EPA, SW 836, 2000. *Methods for Evaluating Solid Waste, Physical/Chemical Methods,*

U. S. Salinity Laboratory Staff. 1954. *Diagnosis and Improvement of Saline and Alkali Soils.* Agriculture Handbook No. 60, U.S. Department of Agriculture (USDA). U.S. Government Printing Office, Washington, DC.

CHAPTER

7

LIMITING CONSTITUENT CRITERIA

7.1 INTRODUCTION

A review of oil and gas drilling and producing activities and analysis of the wastes generated, demonstrated reserve pit solids are not hazardous, dangerous or toxic, (EPA, 1987). These conclusions confirm the industry's own research efforts and study results. By comparing regulations with a demonstrated lack of harmful wastes, EPA concluded E&P activities have little adverse impact on the environment. However, one state, Louisiana (Rule 29-B, 2000), mandates waste treatment protocol and demonstration of treatment effectiveness by comparing postclosure analyses to Louisiana regulatory clean closure values.

In our opinion, most oil and gas wastes require some form of management to preclude adverse impact to land resources, land use or water resources. Hydrocarbons, salts and the dispersed nature of waste solids (lack of physical structure) are the principal problems associated with oil field wastes. Some wastes contain elevated hydrocarbon levels, while others are high in salts. Many oil and gas producing States require reserve pit management and pit closure in an attempt to minimize the potential for an adverse impact. However, the state regulations vary substantially in their requirements and for the most part are ineffective.

7.2 STATE REGULATIONS

Most states either exempt reserve pit construction from regulations or provide demonstration options to operators allowing unlined earthen pit use during drilling. Demonstration may simply take the form of disclosing the anticipated mud program.

Standard operating procedure for pit closure in many states includes dewatering, compacting and backfilling the pit. Some specify the thickness of the final cover and material to be used, usually stripped and segregated topsoil. Liners are required in environmentally (hydrologically) sensitive areas when other than fresh water muds are used. Some states do not require liners, but substitute more stringent timing for liquid removal and closure. All states require closure following cessation of drilling operations unless the pit has been designed and constructed for multiple use. A summary of selected state guidelines or regulations for reserve pit closure is given below:

7.2.1 Alabama

    A. Regulatory Authority/Jurisdiction

        State Oil and Gas Board; Onshore Lands Operation; Chapter 4—Drilling

B. 400-1-4-11 Recycling or Disposal of Pit Fluids and Pit Closure

1. Earthen pits allowed
2. Pit closure. Within 90 days after a well is drilled, completed or worked over all pits must be properly filled and compacted unless otherwise approved by the supervisor. Pits must be backfilled with earth and compacted to the satisfaction of the supervisor. After all fluids and recoverable slurry in the pit is disposed of, the supervisor may permit the operator to leave such pit for use by the landowner, if the surface owner requests in a written statement to the Board that the pit be left open. The written statement should include the intended use for the pit.

7.2.2   Colorado

A. Regulatory Authority/Jurisdiction

Department of Natural Resources; Colorado Oil and Gas Conservation Commission (COGCC)

B. §903. Pit Permitting/Reporting Requirements

1. Earthen pits allowed
    a. No permit is required for fresh water base mud reserve pits, if the concentrations are below the respective value included in (b), below.
    b. Permits are required for reserve pits designed for use with fluids containing hydrocarbon concentrations exceeding 20,000 mg/L TPH or chloride concentrations at total well depth >15,000 mg/L in sensitive areas or >50,000 mg/L outside sensitive areas using Pit Construction Report/Permit, Form 15.

C. §905. Closure of Pits, and Buried or Partially Buried Produced Water Vessels.

Operators of lined pits and buried or partially buried produced water vessels must ensure soils and ground water meet the allowable concentrations of Table 910.1 from COGCC Rule 910.

TABLE 910.1
ALLOWABLE CONCENTRATIONS AND LEVELS

Contaminant of Concern	Allowable Concentrations
Organics in Soil: EPA Method 8015 (modified)	
TPH—Nonsensitive area	10,000 mg/kg
TPH—Sensitive area	1000 mg/kg
Organics in Ground Water: EPA Method 8020[a]	
Benzene	5 ug/L[a]
Toluene	1,000 ug/L[a]
Ethylbenzene	680 ug/L[a]
Xylenes	1400–10,000 ug/L[b]
Inorganics in Soils[d]	
Electrical conductivity (EC)	4 mmhos/cm or 2 times background
Sodium adsorption ratio (SAR)	<12
pH	6–9 s.u.
Inorganics in ground water	
Total dissolved solids (TDS)	<1.25 × background[a]
Chlorides	<1.25 × background[a]
Sulfates	<1.25 × background[a]

(Continued)

TABLE 910.1
(Continued)

Contaminant of Concern	Allowable Concentrations
Total metals in soils: EPA Method 3050[d]	
Arsenic	41 mg/kg[b]
Barium (LDNR [e] true total barium)	180,000 mg/kg[b]
Boron (hot water soluble)	2 mg/L[b]
Cadmium	26 mg/kg[b]
Chromium	1500 mg/kg[b]
Copper	750 mg/kg[b]
Lead	300 mg/kg[b]
Mercury	17 mg/kg[b]
Molybdenum	[c]
Nickel	210 mg/kg[b]
Selenium	[c]
Silver	100 mg/kg[b]
Zinc	1400 mg/kg[b]

[a]Concentrations taken from CDPHE-WQCC (Colorado Department of Public Health and Environment-Water Quality Control Commission.
[b]Concentrations taken from API Metals Guidance: Maximum Soil Concentrations (API Publication Number 4600).
[c]Concentrations are dependent on site-specific conditions.
[d]Consideration will be given to background levels.
[e]LDNR refers to Louisiana Department of Natural Resources.

Pit evacuation. Prior to backfilling and site reclamation, E&P waste must be treated or disposed in accordance with Rule 907.

D. §907. Management of E&P Wastes

Drilling fluids may be recycled, injected into Class II wells, taken to commercial landfill or land treated.

7.2.3 Louisiana

7.2.3.1 Regulatory Authority/Jurisdiction. Department of Natural Resources; Part 19—Office of Conservation; General Operations; Chapter 3—Pollution Control; Onsite Storage; Treatment and Disposal of Nonhazardous Oilfield Waste (NOW) Generated from the Drilling and Production of Oil and Gas Wells (Oilfield Pit Regulations)

A. Pit Construction, Management and Closure

1. Earthen pits allowed, but they must be protected from surface waters by levees or walls and by drainage ditches, where needed, and no siphons or openings will be placed in or over levees or walls which would permit escaping of contents so as to cause pollution or contamination.
2. Liquid levels in pits must not rise within 2 ft of top of pit levees or walls.

B. §311. Pit Closure

1. Reserve pits must be closed to assure protection of soil, surface water, groundwater aquifers and drinking water. Operators may close pits utilizing onsite land treatment, burial, solidification or other techniques approved by the Office of Conservation.

2. Liability for pit closure must not be transferred from an operator to the owner of the surface land(s) on which a pit is located.

3. For evaluation purposes *prior* to closure of any pit and for all closure and onsite and offsite disposal techniques, excluding subsurface injection of reserve pit fluids, nonhazardous oilfield waste (pit contents) must be analyzed for the following parameters:
   a. pH;
   b. Total metals content (mg/kg) for:
      1. Arsenic
      2. Barium
      3. Cadmium
      4. Chromium
      5. Lead
      6. Mercury
      7. Selenium
      8. Silver
      9. Zinc

   c. Oil and grease (percent dry weight)
   d. Soluble salts and cationic distributions:
      1. Electrical conductivity—SPEC in mmhos/cm (millimhos);
      2. Sodium adsorption ratio—SAR;
      3. Exchangeable sodium%age—ESP (percent); and
      4. Cation exchange capacity—CEC (meq/100 g soil).

4. Laboratory Procedures for Nonhazardous Oilfield Waste Analyses
   a. For soluble salts, cationic distributions, metals (except barium) and oil and grease (organics) samples are to be analyzed using standard soil testing procedures as presented in the Laboratory Manual for the Analysis of Oilfield Waste (State wide Order 29-B, LAC 43XIX. Chapters 3, 4 and 5. Department of Natural Resources, August 9, 1988, or latest revision).
   b. For barium analysis, samples are to be digested in accordance with the "True Total" method, as presented in the Laboratory Procedures Manual for the Analysis of Oilfield Waste (Department of Natural Resources, August 9, 1988 or latest revision).

5. Reserve pits utilized in the drilling of wells less than 5000 feet in depth are exempt from the testing requirements of §311.C and §313 provided the following conditions are met:
   a. The well is drilled using only freshwater "native" mud which contains no more than 25 lb/bbl bentonite, 0.5 lb/bbl caustic soda or lime, and 50 lb/bbl barite; and
   b. Documentation of the above condition is maintained in the operator's files for at least three years after completion of pit closure activities.

C. §313. Pit Closure Techniques and Onsite Disposal of NOW

1. Reserve pit fluids, as well as drilling muds, cuttings, etc. from holding tanks, may be disposed of onsite. All NOW must be either disposed of on-site on a one-time basis or transported to an approved commercial facility or transfer station.
2. Prior to conducting onsite pit closure activities, an operator must make a determination the pit closure requirements are attainable.
3. For all pit closure techniques, except solidification, *waste/soil mixtures* must not exceed the following criteria:
   a. Range of pH: 6–9
   b. Total metals content (mg/kg)

Parameter	Limitation (mg/kg)
Arsenic	10
Barium	
Submerged wetlands area	20,000
Elevated wetlands	20,000
Uplands	40,000
Cadmium	10
Chromium	500
Lead	500
Mercury	10.
Selenium	10
Silver	200
Zinc	500

4. Land Treatment. Pits containing NOW may be closed onsite by mixing drilling wastes with soil from pit levees or walls and adjacent areas provided waste/soil mixtures at completion of closure operations do not exceed the above criteria, as applicable.

   a. In addition to the pH and metals criteria listed in §313, above, land treatment of NOW in submerged wetland, elevated wetland and upland areas is permitted if the oil and grease content of the waste/soil mixture after closure is <1% (dry weight).

   b. Additional parameters for land treatment of NOW in *elevated, freshwater wetland* areas where the disposal site is not normally inundated:
      1. Electrical conductivity (EC—solution phase): < 8 mmhos/cm;
      2. Sodium adsorption ratio (SAR—solution phase): <14;
      3. Exchangeable sodium%age (ESP—solid phase): < 25%.

   c. Additional parameters for land treatment of NOW in *upland areas*:
      1. Electrical conductivity (EC-solution phase): <4 mmhos/cm;
      2. Sodium adsorption ratio (SAR—solution phase): <12;
      3. Exchangeable sodium percentage (ESP—solid phase): <15%.

5. Burial or Trenching. Pits containing NOW may be closed by mixing the waste with soil and burying the mixture onsite. Place the bottom of the waste at least 5 ft above the seasonal high water table.

6. Land Treatment. Pits containing NOW may be closed onsite by mixing wastes with soil from pit levees or walls and adjacent areas provided waste/soil mixtures at completion of closure operations do not exceed the above criteria, as applicable, unless the operator can show that higher limits for EC, SAR and ESP can be justified for future land use or that background analyses indicate that native soil conditions exceed the criteria.

   a. In addition to the pH and metals criteria listed in §313.C, above, land treatment of NOW in submerged wetland, elevated wetland, and upland areas is permitted if the oil and grease content of the waste/soil mixture after closure is <1% (dry weight).

   b. Additional parameters for land treatment of NOW in elevated, freshwater wetland areas where the disposal site is not normally inundated:
      1. Electrical conductivity (EC—solution phase): <8 mmhos/cm;
      2. Sodium adsorption ratio (SAR—solution phase): <14;
      3. Exchangeable sodium percentage (ESP-solid phase): <25%.

   c. Additional parameters for land treatment of NOW in upland areas:
      1. Electrical conductivity (EC-solution phase): <4 mmhos/cm;
      2. Sodium adsorption ratio (SAR-solution phase): <12;
      3. Exchangeable sodium%age (ESP-solid phase): <15%.

7.2.4 New Mexico

A. Regulatory Authority/Jurisdiction

New Mexico Energy, Minerals and Natural Resources Department; Oil Conservation Division; Title 19—Natural Resources and Wildlife; Chapter 15—Oil and Gas; Part 17— Pits, Closed-Loop Systems, Below Grade Tanks and Sumps has established new provisions and siting requirements effective June 16, 2008.

B. Design, Construction, and Operational Standards

1. Permit required for pits, below-grade tanks, closed-loop systems, or approved alternatives.
2. All temporary pits must be lined.
3. No pit, below-grade tank, or on-site closure method allowed where ground water is <50 feet below the bottom of the design.
4. On-Site Closure Methods.
   a. In-place burial: Where ground water is between 50 and 100, mix ratio no greater than 3:1, may not exceed closure standards.
   b. On-site trench burial: Where ground water is more than 100 feet below the bottom of the buried waste, mix ratio no greater than 3:1, may not exceed closure standards.
5. Operational Requirements
   a. Operators must notify division if liner damaged and must repair or replace.
   b. Must meet technical standards for temporary pits regarding 2 foot free board and daily inspections.
   c. Must remove drilling fluids within 30 days after the rig is released.

C. Closure Requirements

1. Temporary Pits
   a. Three methods include waste excavation and removal, on-site burial and NMOCD approved alternative.
   b. Waste excavation and removal requires removal of pit contents and liner. Also requires demonstration that soils under liner meet closure standards
      - Where ground water is between 50 and 100 feet, the operator shall collect a five point composite sample to demonstrate benzene does not exceed 0.2 mg/kg; total BTEX does not exceed 50 mg/kg; TPH does not exceed 2500 mg/kg; the GRO and DRO combined fraction does not exceed 500 mg/kg; and chlorides do not exceed 500 mg/kg or the background concentration whichever is greater.
      - Where ground water is > 100 feet, the operator shall collect a five point composite sample to demonstrate benzene does not exceed 0.2 mg/kg; total BTEX does not exceed 50 mg/kg; TPH does not exceed 2500 mg/kg; the GRO and DRO combined fraction does not exceed 500 mg/kg; and chlorides do not exceed 1000 mg/kg or the background concentration whichever is greater.
   c. On-Site burial closure requires compliance with siting criteria, proof of surface owner notification, file deed notice of exact location, and must place a steel marker at the center of burial location. Buried wastes must not exceed the 3:1 mixing ratio limit.
      - Where ground water is 50 to 100 feet below the bottom of the buried waste, the operator shall collect a five point composite sample of the mix and demonstrate that benzene does not exceed 0.2 mg/kg; total BTEX does not exceed 50 mg/kg; TPH does not exceed 2500 mg/kg; the GRO and DRO combined fraction does not exceed 500 mg/kg; and chlorides do not exceed 500 mg/kg or the background concentration whichever is greater.

- Where ground water is > 100 feet below the bottom of the buried waste, the operator shall collect a five point composite sample of the mix and demonstrate that benzene does not exceed 0.2 mg/kg; total BTEX does not exceed 50 mg/kg; TPH does not exceed 2500 mg/kg; the GRO and DRO combined fraction does not exceed 500 mg/kg; and chlorides do not exceed 1000 mg/kg or the background concentration whichever is greater.

    d. On-Site trench burial closure requires compliance with siting criteria and ground water > 100 feet below the bottom of the buried waste. Trench must be lined and buried waste must not exceed the 3:1 mixing ratio limit.
- The operator shall collect a five point composite sample of the mix and demonstrate that TPH does not exceed 2500 mg/kg; and chlorides do not exceed 250 mg/liter for EPA Method 1312 or other EPA leaching procedure approved by the division.
- The operator shall install a geomembrane cover over the filled on-site trench; install the prescribed cover, contour and re-vegetate.

7.2.5   North Dakota

  A. Regulatory Authority/Jurisdiction

    1. Title 43 Industrial Commission of North Dakota; Article 02—Mineral Exploration and Development; Chapter 03—Oil and Gas Conservation; Section 19—Reserve Pit for Drilling Mud and Drill Cuttings—Reclamation of Surface, of the North Dakota Administrative Code.
    2. Division: Bureau of Land Management
    3. Bureau of Land Management
    4. U.S. Forest Service

  B. Pit Construction/Management

    1. Earthen pit allowed
    2. Earthen pits. All earthen pits used during the drilling of a well must be filled and leveled within a reasonable time after the completion of the well.

  C. Reclamation of Surface

    1. In the construction of a drill site, access road, and all associated facilities, the topsoil must be removed, stockpiled and stabilized or otherwise reserved for use when the area is reclaimed.
    2. Within a reasonable time, but not more than 1 year after the completion of a well, the reserve pit must be reclaimed.

7.2.6   Oklahoma

  A. Regulatory Authority/Jurisdiction

    Title 165, Oklahoma Corporation Commission, Chapter 10—Oil and Gas Conservation; Regulated under Oklahoma Rules 165:10-1-2, 165:10-7-16, 165:10-7-19 and Appendix I of rule.

  B. Pit Construction/Management

    1. Earthen pit are allowed.
    2. Reserve pits. The operator of the pit must indicate the type of mud system(s) to be used, the maximum and average anticipated chloride concentration of the mud (based on drilling records in the area), and whether or not pit fluids will be segregated.

3. Liner requirements. The Commission's Technical Department will indicate whether or not a liner is required.
4. Construction requirements
   a. Stockpile of topsoil.
   b. Exclude of runoff water.

5. Operation and maintenance requirements.
   a. Freeboard. The fluid level of any noncommercial pit must be maintained at all times at least 24 in. below the lowest elevation on the top of the berm.
   b. Reserve/circulation pits. The operator of any reserve/circulation pit must limit its contents to the fluids and cuttings to a single well.

6. Closure requirements
   a. Designation of disposal method. The operator of any reserve/circulation pit must indicate the proposed method of disposal of drilling fluids and/or cuttings. Options must be limited to the following, unless written a Field Operations representative grants approval:
      1. Evaporation/dewatering and backfilling.
      2. Chemical solidification of pit contents.
      3. Annular injection (requires permit).
      4. Land application (requires permit).
      5. Disposal in permitted commercial pit.
      6. Disposal at permitted commercial soil farming facility.
      7. Disposal at permitted recycling/reuse facility.

   b. One-time land application of water base fluids from earthen pits allowed
      1. Site suitability
      2. Soil depth; 20 in. or more to bedrock
      3. Soil texture; at least 12 in. cumulative or individually of loam, silt loam, silt, sandy clay loam, silty clay loam, clay loam, sandy clay, silty clay or clay
      4. Soil salinity
         a. EC <4 mmhos/cm in upper 6 in. of soil (by SPEC or 1:1 by weight)
         b. ESP <10%
         c. TDS = EC × 640 mg/L

      5. Depth to water table. >6 ft from soil surface
      6. Sampling requirements,
         a. Notice to district—Notify field inspector at least 2 days before sampling
         b. Receiving soil —four samples (0–6 in. deep) and four samples (12–20 in. deep) per 10 acres, or part thereof, composited into 2-pint samples by depth

      7. Drilling fluids/cuttings
         a. No freshwater, except natural precipitation, should be added to reserve pit
         b. Five samples per 25,000 bbl or less, plus one sample per 5000 bbl over 25,000 bbl

      8. Receiving soil analysis
         a. Electrical conductivity (SPEC)
         b. Exchangeable sodium percentage (ESP)

      9. Drilling fluid/cuttings analyses
         a. Electrical conductivity (EC)
         b. Oil and grease (O&G)
         c. Solids dry weight
     10. Loading calculations. Refer to loading worksheet below.

11. Land application method
    a. Apply about 4 in. thick
    b. If more than 500 lb/acre of oil and grease and/or 50,000 lb of solids are applied, disk and cross-disk to 6 in. deep
C. Work Sheet

7.2.7 Total Dissolved Solids (TDS)

For solid material:
EC of receiving soil_____mmhos/cm × 640[a] =_____mg/L TDS in receiving soil.
TDS in receiving soil_____mg/L × 2 =_____lb/acre TDS in receiving soil.
6000 lb/ac TDS −_____lb/ac TDS in receiving soil = _____*maximum TDS* (lb/ac) to be applied.
EC of materials to be applied _____mmhos/cm × 640[a] = _____mg/L *TDS.*
Maximum TDS (lb/acre) to be applied_____ ÷ (TDS of materials to be applied_____mg/L × 0.000001) = maximum weight of materials to be applied _____lb/acre

For liquid materials:
Maximum weight of materials to be applied _____lb/acre ÷ (sample weight _____ lb/gal × 42) = *maximum loading _____ bbl/acre.*
Total volume of materials to be applied _____bbls ÷ *maximum loading _____ bbl/acre = minimum acres* required _____.

For solid materials:
Maximum weight of materials to be applied _____lb/acre ÷ (sample weight _____lb/gal[1] × 202) = *Maximum* loading _____ cu yd/ac.
Total volume of materials to be applied_____cu yd ÷ *maximum loading*_____cu yd/acre = *Minimum acres* required

7.2.7.1 Chlorides (Cl)

Cl in receiving soil_____ mg/kg × 2 = _____lb/acre Cl in receiving soil.
3500 lb/ac Cl −_____lb/ac Cl in receiving soil = *maximum Cl* (lb/acre) to be applied _____.
Maximum Cl (lb/ac) to be applied _____ ÷ Cl in the material to be applied _____ mg/kg × 0.000001 = *maximum weight* of materials to be applied _____ lb/ac

For liquid materials:
Maximum weight of materials to be applied _____ lb/acre ÷ (sample weight _____ lb/gal × 42 gal/bbl) = *maximum loading* _____ bbl/acre.
Total volume of materials to be applied_____ bbl ÷ *Maximum loading* _____ bbl/acre = *Minimum acres* required _____.

For solid materials:
Maximum weight of materials to be applied_____ lb/acre ÷ (sample weight_____lb/gal × 202) = *Maximum loading*_____cu yd/ac.
Total volume of materials to be applied _____ cu yd ÷ *maximum loading* _____cu yd/acre = *minimum acres* required_____.

7.2.7.2 Oil and Grease (O&G)

40,000 lb/acre O&G ÷ (O&G of materials to be applied _____ mg/L × 0.000001) = *maximum weight* of materials to be applied . _____ lb/acre

_____
[1]Based on dry weight percentage of composite sample of materials. (Source: Added at 12 OK Reg. 2039, 7-1-95).
[a]Constant given in USDA Handbook 60 to convert EC to TDS.

For liquid materials:
Maximum weight of materials to be applied _____ lb/acre ÷ (sample weight _____ lb/gal × 42 gal/bbl) = *maximum loading_____bbl/acre.*
Total volume of materials to be applied _____bbl ÷ *maximum loading_____ bbl/acre = minimum acres* required

For solid materials:
Maximum weight of materials to be applied _____ lb/acre ÷ (sample weight _____lb/gal ×202) = *maximum loading_____ cu yd/acre.*
Total volume of materials to be applied_____cu yd ÷ *Maximum loading_____cu yd/acre = Minimum acres* required _____.

Dry Weight:
Wet weight of drilling mud_____lb/gal × _____% dry solids = _____lb/gal dry weight ____ ____lb/gal dry weight × 202 = _____lb/cu yd
200,000 lb/acre dry weight ÷ _____lb/cu yd = *maximum cu yd/acre*_____.
Total volume of materials to be applied _____cu yd ÷ *maximum cu* yd/acre_____ = *Minimum acres* required_____.

7.2.8   Texas: Pits

A. Regulatory Authority/Jurisdiction

Title 16—Economic Regulation; Part 1—Railroad Commission of Texas; Title 003—Oil & Gas Division; Chapter 3—Oil and Gas Division

B. Pit Construction/Management—§3.8
Rule 8. Pollution control.

1. Earthen pits allowed
    a. Drilling fluids, whether fresh water base, saltwater base, or oil base;
    b. Drill cuttings, sands and silts separated from the circulating drilling fluids;
    c. Wash water used for cleaning drill pipe and other equipment at the well site;
    d. Drill stem test fluids; and
    e. Blowout preventer test fluids.

2. Authorized disposal methods
    a. Low-chloride drilling fluid. A person may, without a permit, dispose of the following oil and gas wastes by land treating, provided the wastes are disposed of on the same lease where they are generated, and provided the person has the written permission of the surface owner of the tract where land treating will occur: water base drilling fluids with a chloride concentration of 3000 mg/Lor less; drill cuttings, sands and silts obtained while using water base drilling fluids with a chloride concentration of 3000 mg/L or less; and wash water used for cleaning drill pipe and other equipment at the well site.
    b. Other drilling fluid. A person may, without a permit, dispose of the following oil and gas wastes by burial, provided the wastes are disposed of at the same well site where they are generated: water base drilling fluid which had a chloride concentration in excess of 3000 mg/L but which have been dewatered; drill cuttings, sands, and silts obtained while using oil base drilling fluids or water base drilling fluids with a chloride concentration in excess of 3000 mg/L; and those drilling fluids and wastes allowed to be land treat without a permit.

3. Backfill requirements.
   a. A person who maintains or uses a reserve pit, mud circulation pit, fresh makeup water pit, fresh mining water pit, completion/workover pit, basic sediment pit, flare pit or water condensate pit must dewater, backfill, and compact the pit according to the following schedule.
      1. Reserve pits and mud circulation pits which contain fluids with a chloride concentration of 6100 mg/L or less and fresh makeup water pits must be dewatered, backfilled and compacted within 1 year of cessation of drilling operations.
      2. Reserve pits and mud circulation pits that contain fluids with a chloride concentration in excess of 6100 mg/L must be dewatered within 30 days and backfilled and compacted within 1 year of cessation of drilling operations.
      3. If a person constructs a sectioned reserve pit, each section of the pit must be considered a separate pit for determining when a particular section should be dewatered.

   b. A person who maintains or uses a reserve pit, mud circulation pit, fresh makeup water pit or completion/workover pit must remain responsible for dewatering, backfilling, and compacting the pit within the time prescribed by clause (1), above, even if the time allowed for backfilling the pit extends beyond the expiration date or transfer date of the lease covering the land where the pit is located.

C. Soil/Waste Contaminated by Hydrocarbons—§3.91

Rule 91—Cleanup of Soil Contaminated by a Crude Oil Spill

7.2.9 Texas: Soils

A. Cleanup of Soil Contaminated by a Crude Oil Spill—§3.91 (Rule 91)

   1. Scope. These cleanup standards and procedures apply to the cleanup of soil in non-sensitive areas contaminated by crude oil spills from activities associated with the exploration, development, and production, including transportation, of oil or gas or geothermal resources. For the purposes of this section, crude oil does not include hydrocarbon condensate. These standards and procedures do not apply to hydrocarbon condensate spills, crude oil spills in sensitive areas, or crude oil spills that occurred prior to the effective date of this section. Cleanup requirements for hydrocarbon condensate spills and crude oil spills in sensitive areas will be determined on a case-by-case basis. Cleanup requirements for crude oil contamination that occurred wholly or partially prior to the effective date of this section will also be determined on a case-by-case basis. Where cleanup requirements are to be determined on a case-by-case basis, the operator must consult with the appropriate district office on proper cleanup standards and methods, reporting requirements, or other special procedures. The effective date is January 1, 1994.

   2. Terms. The following words and terms, when used in this section, must have the following meanings, unless the context clearly indicates otherwise.
      a. Sensitive areas—These areas are defined by the presence of factors, whether one or more, that make an area vulnerable to pollution from crude oil spills. Factors that are characteristic of sensitive areas include the presence of shallow groundwater or pathways for communication with deeper groundwater; proximity to surface water, including lakes, rivers, streams, dry or flowing creeks, irrigation canals, stock tanks and wetlands; proximity to natural wildlife refuges or parks; or proximity to commercial or residential areas.
      b. Requirements for cleanup.
         1. Removal of free oil. To minimize the depth of oil penetration, all free oil must be removed immediately for reclamation or disposal.

2. Delineation. Once all free oil has been removed, the area of contamination must be immediately delineated, both vertically and horizontally. For purposes of this paragraph, the area of contamination means the affected area with >1.0% by weight total petroleum hydrocarbons.
3. Excavation. At a minimum, all soil containing >1.0% by weight total petroleum hydrocarbons must be brought to the surface for disposal or remediation.
4. Prevention of storm water contamination. To prevent storm water contamination, soil excavated from the spill site containing >5.0% by weight total petroleum hydrocarbons must immediately be:
   a. Mixed in place to 5.0% by weight or less total petroleum hydrocarbons; or
   b. Removed to an approved disposal site; or
   c. Removed to a secure interim storage location for future remediation or disposal. The secure interim storage location may be on-site or off-site. The storage location must be designed to prevent pollution from contaminated storm water runoff. Placing oily soil on plastic and covering it with plastic is one acceptable means to prevent storm water contamination; however, other methods may be used if adequate to prevent pollution from storm water runoff.

3. Remediation of soil.
   a. Final cleanup level. A final cleanup level ≤1.0% by weight total petroleum hydrocarbons must be achieved as soon as technically feasible, but not later than one year after the spill incident. The operator may select any technically sound method that achieves the final result.
   b. Requirements for bioremediation. If on-site bioremediation or enhanced bioremediation is chosen as the remediation method, the impacted soil to be bioremediated must be mixed with ambient or other soil to achieve a uniform mixture that is no more than 18 inches in depth and contains no more than 5.0% by weight total petroleum hydrocarbons.

7.2.10 Utah

A. Regulatory Authority

Utah (1996 Guidelines); Title R649—Oil, Gas and Mining: Oil and Gas, Environmental Handbook, "Environmental Regulations for the Oil & Gas Exploration and Production Industry" Prepared by G.L. Hunt, Environmental Manager, Utah Division of Oil, Gas & Mining, January 1996.

B. Guidance for Reserve Pit Closure

Recommended procedures to be followed when closing reserve pits used during the drilling of oil and/or gas wells and other exploratory test holes.

1. Procedures:
   A reserve pit should be closed within 1 year following drilling and completion of a well. A pit is considered cleaned up when it meets the following recommended levels. Operators should avoid putting wastes other than drill cuttings, mud and completion fluids into a reserve pit as such a practice could complicate pit closure requirements. Liquid in a pit should be allowed to either evaporate or be removed. If removed, it must be disposed of properly, some options are injection (in this well or another), hauled to a permitted disposal facility, or reused at another well. Pit liners can be cut off above the cuttings/mud level and hauled to a landfill, or folded in and processed along with other pit contents and covered. No remnants of liner material can be exposed at the surface when pit closure is

complete. Pit area should be mounded so as not to allow ponding of water and drainage diverted around as not to allow erosion of the old pit site.

2. Backfill Closure:

   For fresh water mud pit, where a liner is not required, and pit does not contain much oil (TPH $\leq$ 3 wt%). The pit can simply be backfilled after the fluids are removed, evaporated and/or percolated.

3. Dilution Burial:

   This method should not be used in general if the water table is less than 10 feet below the pit bottom, especially if intervening soil is permeable, for example, sandy or gravelly soils. The pit contents are mixed with adjacent soil to reduce constituents levels below recommended levels (SPEC $\leq$ 12 mmhos/cm, TPH $\leq$ 3%), or higher if background levels are higher or with Division approval. After mixing, the pit contents should contain no more than about 50% moisture by weight prior to burial of a waste/soil mix. Mixed contents should be covered with at least 2 ft of soil including topsoil, if possible.

4. Solidification:

   This method commonly uses cementitious/pozzolanic processes that envelope the waste solids in a materials matrix. The mixed pit contents should be covered with at least 2 ft of soil including topsoil, if possible.

5. Land Treatment:

   Pit contents (after fluid removal) can be spread over a location and mixed in, if cleanup levels are met as determined using the Division's guidance for estimating cleanup levels for petroleum-contaminated soils. A pit can then be backfilled.

6. Impact Cleanup Levels:

   Utah developed a Ranking System for determining clean-up levels for salts and hydrocarbons. As of December 2007 the cleanup levels were being revised by the State. Check with the Utah Division before using these values in Utah. The current recommendations are available at http://ogm.utah.gov/oilgas/PUBLICATIONS/Handbooks/envbook. htm#cleanup

7.2.11 Wyoming

A. Regulatory Authority/Jurisdiction

   1. Office of Lands and Investments; Oil & Gas Conservation Commission; Chapter 4— Environmental Rules, Including Underground Injection Control Program Rules for Enhanced Recovery and Disposal Projects
   2. Bureau of land Management

B. Pollution and Surface Damage

   1. The Commission exercises its regulatory authority over the construction, location, operation and reclamation of oilfield pits. The following pits are subject to this regulation:
      a. Reserve pits on the drilling location;
      b. Reserve pits off the location within a lease, unit or communitized area permitted by owner or unit operator drilling the well;
      c. In addition, the earthen pit allowed.

   2. Pit Closure. Pit may be closed using the usual method of onsite natural evaporation and subsequent burial of solids. Trenching or squeezing (dilution burial) of pits is expressly prohibited. Burial methods cannot compromise the integrity of manufactured, soil mixture, or recompacted clay liners without written approval by the Supervisor. One-time land spreading of reserve pit fluids on the *drilling pad* may be approved upon submittal

of analyses, mud recaps and treating summaries, groundwater identification, and other information that is deemed appropriate. Prior approval must be obtained from the Supervisor, if drilling fluid is disposed on the drill pad. Based on site-specific conditions, the Supervisor will determine closure standards and testing requirements for all pits.

3. Special Requirements. Pit solids showing high concentrations of salt (exchangeable sodium >15%) must be removed from the location and disposed in a permitted facility, encapsulated or chemically or mechanically treated. Oil-based mud solids must be removed and disposed in a permitted facility, or mixed with soil to less than 1% oil and grease content by weight at burial, solidified using a Commission-approved commercial pit treatment or road spread or land spread or land treated in accordance with Commission or Department of Environmental Quality (DEQ) rules. DEQ and the Commission share jurisdiction over road spreading or road application. Burial after encapsulation or solidification will be approved, if the stabilized mixture contains less than 10 mg/L leachable oil.

4. Land treatment must be approved by the DEQ.

5. Replace Topsoil. Backfill or grade, and replace topsoil, or approved subsoil, which has been segregated and preserved as may be required in the approved reclamation plan.

7.3   AMERICAN PETROLEUM INSTITUTE METALS CRITERIA

The American Petroleum Institute (API) published (API, 1995) metals criteria for E&P wastes, Table 7.1.

Molybdenum: On February 25, 1994, (59 FR 9095), EPA rescinded the risk-based maximum soil concentration for Mo of 9 mg/kg due to technical errors in 40 CFR §503.13 and established a non-risk-based interim ceiling limit of 75 mg/kg (58 FR 9387, February 19, 1993). Under conditions of alkaline soils and arid or semiarid conditions with deficient levels of copper in the soil this interim level may not be protective of grazing livestock.

TABLE 7.1
API RECOMMENDED E&P WASTE METALS CRITERIA

Metal	Extraction Method	Maximum Soil Concentration, mg/kg
Arsenic	EPA method 3050B[a]	41
Barium	LDNR true total barium[b]	180,000
Boron	Hot water soluble[c]	2 mg/l[d]
Cadmium	EPA method 3050B[a]	26
Chromium	EPA method 3050B[a]	1,500
Copper	EPA method 3050B[a]	750
Lead	EPA method 3050B[a]	300
Mercury	EPA method 3050B[a]	17
Molybdenum	EPA method 3050B[a]	See below[e]
Nickel	EPA method 3050B[a]	210
Selenium	EPA method 3050B[a]	See below[e]
Zinc	EPA method 3050B[a]	1400

[a]EPA, 1986. Testing Methods for Evaluating Solid Waste. SW-846 3rd ed.
[b]Louisiana Department of Natural Resources, 1988. Laboratory Procedures for Analysis of Oilfield Waste, Statewide Order 29-B.
[c]Carter, M.R. 1993. Soil Sampling and Methods of Analysis. Lewis Publishers, Boca Raton, FL, pp. 91–93.
[d]Units in mg/L soluble rather than mg/kg total.
[e]Plant availability and risk to grazing livestock dependent on soil pH.

Selenium: EPA, using the risk-based multipathway analysis, generated the limiting pathway concentration of 100 mg/kg (see Table 3, 40 CFR §503.13). However, the potential for plant uptake of Se may be high in alkaline soils under arid and semiarid conditions. Plants accumulating Se under these soils conditions may pose a threat to grazing animals. Therefore, if elevated levels of Se are found in the waste, the operator should consider site conditions that control its availability.

7.4   PIT UTILIZATION AND WASTE MANAGEMENT ("NOW" DEFINED)

Nonhazardous Oilfield Waste (NOW) is defined as exempted liquids or solids generated in the exploration and production of oil and gas as listed below. These wastes were exempted by the 1980 amendments to the Resource Conservation and Recovery Act (RCRA), which regulates E&P wastes under RCRA Subtitle D. EPA recommended the continuance of the NOW exemption from Subtitle C of RCRA in the December 1987 report to Congress. Based on a more recent EPA study, NOW most likely will remain exempt from RCRA Subtitle C regulations.

Summary listing of nonhazardous oilfield wastes (NOW)

1. Produced brine or produced water
2. Oil base drilling mud and cuttings
3. Water base drilling mud and cuttings
4. Drilling, workover and completion fluids
5. Production storage tank sludges
6. Produced oily sands and solids
7. Produced formation fresh water
8. Rainwater from ring levees and impoundments at production and drilling facilities
9. Washout water generated from the cleaning of vessels (barges, tanks, etc.) used to transport NOW and are not contaminated by hazardous waste or material
10. Washout pit water from oilfield related carriers not permitted to haul hazardous waste or materials
11. Nonhazardous natural gas plant processing waste, which is or may be commingled with produced formation water
12. Waste from approved salvage oil operators that only receive waste oil (basic sediment and water, BS&W) from oil and gas leases
13. Pipeline test water that does not meet discharge limitations established by the appropriate state agency, or pipeline pig water
14. Wastes transferred between permitted commercial facilities
15. Material used in crude oil spill cleanup operations. State regulations are based on the assumption that oilfield wastes are nonhazardous and may be treated on-site with the landowner's permission. Treatment usually effects waste disposal and pit closure.
16. Salvageable hydrocarbons bound for a permitted salvageable oil operator

7.5   SITE INDEXING

On-site waste disposal management is predicated on impacted land resource; waste constituent(s) requiring treatment, and intended land-use. Land resource designations include *brackish wetlands, freshwater wetland and upland sites.*

Further, upland sites are delineated according to intended land-use, for example, woodland/recreational and residential/farming. Depth to groundwater is an important parameter at upland sites where dilution burial is a disposal option. Wetland site indexing simply entails a landform designation followed by a descriptive modifier of the adjacent or encompassing water body (that is, fresh or brackish/saline).

TABLE 7.2
GLC VALUES UTILIZING LAND TREATMENT AND VERTICAL DILUTION

Parameter	Brackish	Wetlands Freshwater Submerged	Wetlands Freshwater Elevated	Uplands Forest/ Recreation	Uplands Farming/ Residential
pH, s.u.	6-9	6-9	6-9	6-8	6-8
SPEC, mmhos/cm	NA[a]	NA	8–50[b]	4–8	2–4
SAR, unitless	NA	NA	14–100[b]	<14	<14
CEC, mEq/100g	>15	>15	>15	>15	>15
ESP, %	NA	NA	<25	<15	<15

Total Metals, mg/kg	Wetlands	Uplands
Arsenic	10	40
True total barium	20,000	40,000
Cadmium	10	10
Chromium	500	500
Lead	500	500
Mercury	10	10
Selenium	10	10
Silver	10	10
Zinc	500	500

Petroleum hydrocarbon, %	Wetlands	Uplands
Oil & Grease/TPH	1	1

[a]NA, not applicable.
[b]Acceptable upper range determined by background value.

7.6   GUIDELINES FOR LIMITING CONSTITUENTS

Guidelines for limiting constituents (GLC) consist of *after treatment* criteria defining the acceptable upper thresholds for waste parameters of interest: i.e., salt, heavy metals and hydrocarbons. Those parameters exceeding limiting constituent levels, generally, require some level of treatment for *one-time* on-site management. GLC values provide a guide for evaluating waste materials. Typically, treatment or management of the most limiting characteristic (most excessive parameter) brings the other constituents below GLC values. GLC values for land treatment; impoundments and remote field pits are presented in Table 7.2. Refer to Chapter 8 for more details.

REFERENCES

American Petroleum Institute (API). 1995. "Metals Criteria for Land Management of Exploration and Production Wastes". Technical Support Document for API Recommended Guidance Values. API Publication Number 4600, Washington, DC.

EPA, 1987. "Report to Congress: Management of Wastes from the Exploration, Development and Production of Crude Oil, Natural Gas, and Geothermal Energy". EPA/530-SW-88-003, 3 vols.

Rule 29-B, 2000. Title 43, Part 19, Chapter 3, §§ 311, 313 and 315—Pollution Control, State of Louisiana, Department of Natural Resources, Office of Conservation, Baton Rouge, LA, Promulgated January 20, 1986; revised October 20, 1990 and December 2000.

CHAPTER

8

SOIL AMENDMENTS

8.1 INTRODUCTION

Extensive research sponsored by EPA (1987) and the oil and gas industry demonstrates the vast majority of E&P wastes are not hazardous. However, many E&P wastes require management to prevent adverse impacts to soil and water. Excessive alkalinity, salts and the dispersed nature of drilling waste solids (lack of physical structure) represent the principal limitations associated with disposal/ remediation of these wastes.

8.2 SOIL REACTION (pH) AMENDMENTS

Aluminum sulfate $[Al_2(SO_4)_3]$ additions control excess alkalinity while the *wastes are in the pit.* Excess acidity in the pit is managed with lime (CaO) additions. Amendments applied to waste liquids will not react properly or fully with pit solids unless blended by mechanical action.

At *upland sites*, final pH adjustments of pit solids are made *after* spreading the solids on land surfaces, and mixing with surface soils. Then, agricultural limestone $[Ca(CO_3)]$ is used to adjust excess acidity and elemental sulfur (S) is used to adjust excess alkalinity. At *wetland sites*, final pH adjustments are made after homogenizing pit contents, but prior to mixing of the pit solids with the levee materials.

For details on desirable pH soil reaction targets and the various mechanisms involved refer to Appendix IV.

8.3 SALINITY AMENDMENTS

Only a rinse or leach process with fresh water reduces salts in soils and E&P waste solids. Poor *physical condition and inherent low hydraulic conductivity of salt impacted soils/wastes prevent free exchange of salt without the aid of structuring amendments, wetting agents and/or mechanical processes.*

Structuring amendments aid salt impacted soil/waste by increasing soil/waste aggregation. These agents help prevent dispersion of the clays as the result of saline (produced waters or salt muds) spills by allowing water to infiltrate. This aggregation aids chemical amendments in entering the soil matrix. Also, polymer products increase soil aggregation. These agents include polyvinyl alcohols, polyacrylamides and natural plant polymers.

Mechanical processes physically loosen the soil allowing a greater surface area for soil contact with water and chemical amendments. Disking can loosen the soil to a depth of 6–10 in. depending on the equipment available. Chiseling the soil is necessary on sites where salt contamination occurs at greater depths.

8.3.1 Sodicity Amendments

Sodium chloride (NaCl) is the primary salt associated with oil field wastes or produced waters. Soils, drill cuttings and other E&P waste solids exposed to high salt levels become sodium saturated and/or sodic (ESP >15%). Calcium amendments counteract sodicity and high salinity. If the material is *both saline (SPEC > 4 mmhos/cm and pH<8.3 s.u.) and sodic*, calcium should be introduced *prior* to the leach process. The amendment of choice is dependent upon solid phase calcium levels and available treating time, *however, applying and tilling organic material (hay, rice hulls, etc.) into the waste/soil immediately after a release will reduce the adverse impact of the salt and most likely will eliminate the need chisel plowing.*

8.3.2 Solid Phase Calcium in Excess

Excess solid phase calcium is a condition typified by salt base muds or produced waters spilled or discharged to calcareous soils, which typically results in a sodic soil condition. The source of solid phase calcium is either $CaCO_3$ in calcareous soils or $CaSO_4$ in gypsiferous soils or both. The amendment of choice is elemental sulfur where free $CaCO_3$ is present, when time is not a consideration, and mechanical means are available to incorporate the amendment. The elemental sulfur oxidizes to sulfuric acid in the soil. This reacts with calcium carbonate to form calcium sulfate.

When time is a factor or mechanical means are not available for distribution, soluble calcium sources, such as calcium chloride ($CaCl_2$) or calcium nitrate [$Ca(NO_3)_2$], are necessary. Also, soluble calcium salts are effective when soil or waste solids have not yet equilibrated to a sodic condition, i.e., applied immediately after a spill.

8.3.3 Solid Phase Calcium Limiting

Agricultural grade gypsum is an effective calcium amendment when time is not a factor and mechanical means are available for distribution. Gypsum is readily available at low cost.

8.3.4 Organic Matter Amendments

Organic amendments in the form of peat moss, low-salt manures and/or hay improves soil structure. Also organic materials serve as a sink for salts by exchange mechanisms and dilute salts in soil solution by increasing moisture-holding capacity. The application of organic matter immediately after a release to soil *reduces time and cost of remediation.*

8.3.5 Bioremediation Amendments

Native soils generally contain a sufficient number and species diversity in microflora (bugs) to degrade petroleum hydrocarbon under controlled applications. In rare cases, inoculations may be necessary to offset poor population dynamics. Typically, an inordinately long lag phase results from limited populations. Addition of commercial microbes reduces the lag phase. However, their effectiveness is difficult to measure when compared to regulatory benchmarks such as oil and grease (O&G) and total petroleum hydrocarbons (TPH).

Fertility adjustments in the form of fertilizer amendments, aeration and moisture are necessary for bioremediation progress. Fertilizer amendments include nitrogen (N), phosphorus (P) and potassium (K). Ammonium nitrate, ammonium sulfate and urea are good nitrogen sources. Nitrogen amendments should be applied in multiple applications to increase efficiency, because nitrogen is lost to the atmosphere with time. Nitrogen should be disked into the soil/waste immediately after application to distribute the amendment and aerate the admixed system.

Nitrogen is applied at rate yielding a carbon:nitrogen (C:N) ratio of 150:1. Some references suggest C:N ratios as low as 25:1. Extensive field experience demonstrates field variations dictate a C:N ratio of 150:1 is highly satisfactory. Phosphorus and potassium are applied at 1/4 the rate of nitrogen (600:1). Superphosphate and muriate of potash are suitable sources. Phosphorus and potassium are applied in a single application with the first nitrogen treatment.

Moisture content should be maintained within 50–80% of field capacity (1 in. of water per week) to effect optimum degradation rates.

8.4 E&P RELEASE EVENTS AND TREATMENT SCENARIOS

The following scenarios illustrate the calculations and procedures required to remediate typical oil and gas field impacted soils/wastes.

8.5 SCENARIO I—LIGHT CRUDE OIL RELEASE ON SOIL

Oil released onto native soil, 20 barrels of West Texas crude oil, gravity 20° API.

8.5.1 Considerations

Oil spilled on native soil spreads in the direction of the surface drainage with pooling in various depressions. Vertical infiltration is controlled by:

- Surface condition
- Soil porosity
- Pore size distribution
- Soil moisture
- Soil structure
- Viscosity and gravity of the oil

Permeability of the soil profile, surface soil condition and soil moisture content are the most important factors affecting infiltration of the oil. The amount of oil percolating into a soil profile is controlled by the permeability of the *least* pervious horizon or layer of soil in the vertical cross section. Guidelines proposed by the Texas Railroad Commission (Statewide Rule 91) require removal of all pooled oil and immediate delineation of the area of contamination, both vertically and horizontally. The area of oil contamination means the affected area with >1 wt% total petroleum hydrocarbon (TPH).

Any soil having more than 5% TPH requires removal or reduction to 5% or less by mixing with uncontaminated (background) soil. All impacted soil must be reduced to = 1% TPH within 1 year following the spill event. *Typically, bioremediation is not effective for reducing TPH below 6–10 in. in depth.* Accordingly, by Rule impacted soil below a depth 12 in. must be brought to the surface for treatment.

8.5.2 Site Assessment

Crude oil with an API gravity of 20 has a specific gravity of 0.93, for example, [141.5/(131.5 + 20)]. A 20 bbl spill of this oil yielding an average concentration of 1% oil to a depth of 6 in. will occupy an area of 14,100 ft^2 (0.33 acre-6 in.). The same volume yielding an average concentration of 5% oil in soil affects 2820 ft^2 (approximately 0.06 acre-6 in.).

Stained soil defines the surface boundaries of a recent oil release. Samples collected in vertical profile define concentrations with depth. Unless specific information dictates otherwise, profiles are sampled initially in 0–6, 6–12, 12–24 and 24–36 in. depth intervals. Profiles should be taken at various points along a transect, which defines surface drainage of the release site. Sample the site in grid fashion, where drainage patterns and internal percolation are not well defined.

8.5.3   Remediation Procedure

The following remediation procedure assumes a 5% TPH level in a 6-in. surface layer following the removal of pooled oil. Contaminated soil below the treatment zone is considered to be <1% TPH.

We assume a mean annual soil temperature >22°C. Decrease allowable TPH concentration in soil to be treated, 1% for each 7°C decrease in temperature below 22°C (Table 8.1). Microbial degradation of petroleum hydrocarbon does not occur when soil temperature remain below 5°C. Consider composting operations when soil temperatures remain below 8°C. This involves removing TPH contaminated soil and incorporating it in a pile of organic material such as high-protein hay and animal manure. The heat generated by decomposition of the organic material elevates the treating temperature, so bioremediation occurs.

Remediation entails biodegradation of crude oil using native soil microflora. Parameters having a profound affect on rates of degradation include:

- Initial oil level (loading rate)
- Carbon:nitrogen ratio and other fertilizer inputs
- Moisture content
- Aeration

The above parameters can be controlled by the operator in the field.

Research shows the optimum carbon:nitrogen (C:N) ratio for field degradation of petroleum hydrocarbon is 150:1. A lower C:N ratio (more fertilizer) theoretically will improve degradation rates. Generally some other variable such as water use efficiency or free air exchange limits the utility of adding more nitrogen. Nitrogenous compounds released by Kjeldahl digests add little available nitrogen reserves, and should not be considered in determining the C:N ratio. Nitrate-N and ammonium-N released by extraction with a potassium chloride solution comprise the available nitrogen fraction and must be considered in determining the C:N ratio if the field or spill area has been fertilized recently with nitrogen.

Phosphorus and potassium are elements that may be limiting and cause a positive response when added to the system. A nitrogen: phosphorus: potassium (N:P:K) ratio of 4:1:1 provides a good balance of these elements.

Mechanically till fertilizer applications into the soil to distribute the amendment through the treatment zone. Tilling also infuses oxygen into the soil. Split applications coupled with tillage operations increase fertilizer use efficiency, and lessen the potential for eutrophic stagnation (the creation of an oxygen deficit by nutrient enrichment).

Moisture levels should be maintained between 50 and 80% of the soil water holding capacity (field capacity). This requires addition of about 1 in. of irrigation water per week in dry climates (arid and

TABLE 8.1
INITIAL TPH LOADING RATES FOR OIL;
20° API AND ABOVE

Mean Annual Soil Temperature, °C	Loading Rate, % TPH
≤22.0	5
15.0–21.9	4
8–14.9	3
<8	0

semiarid). Humid regions with spurious rainfall distribution patterns may require seasonal irrigations to overcome drought conditions.

8.5.4   Calculations

Assumptions:

- Soil temperature > 22°C, TPH loading rate = 5% (Table 8.2)
- 1 acre-6 in. = 2,000,000 lb [91.8 (lb./ft^2)]
- mg/kg × 2 = lb/acre-6 in.
- Crude oil contains 78% carbon
- 0.06 acre-6 in. deep (calculated)

1. Carbon content, in mg/kg, of soil resulting from oil   spill

$$C, \% = TPH\% \times 0.78$$
$$C, \% = 5.0\% \times 0.78$$
$$C, \% = 3.9\%$$
$$C, mg/kg = 3.9\% \times 10,000$$
$$= 39,000 \text{ mg/kg}$$

2. Nitrogen requirement

$$C:N = 150:1$$
$$(39,000 \text{ mg/kg})/N = 150:1$$
$$N = 260 \text{ mg/kg}$$
$$N = (260 \text{ mg/kg}) \times 2$$
$$N = 520 \text{ lb/acre-6 in.}$$

3. Spill site occupies 0.06 acre-6 in.

$$N = 520 \text{ lb/acre-6 in.} \times 0.06 \text{ acre-6 in.}$$
$$N = 31 \text{ lb}$$

4. NPK 4:1:1

$$P,K = 31 \text{ lb N}/4$$
$$N/4 = P,K = 8 \text{ lb}$$

8.5.5   Procedure

1. Remove all pooled oil.
2. Excavate contaminated soil below 6 in. (e.g., >1% TPH).
3. Mix in place to 5% by weight TPH. Maintain a depth of mixed soil × 6 in.
4. Add half of the ammonium nitrate, 47 lb (33% N), 87 lb superphosphate (9.2% P) and 16 lb muriate of potash (51% K).
5. Disk and cross-disk to 6 inches to distribute fertilizer and aerate treatment zone.
6. After 6 to 8 weeks add half of the remaining ammonium nitrate, 23 lb, disk and cross-disk to distribute fertilizer and aerate treatment zone.
7. After 6–8 weeks add the remaining ammonium nitrate, 23 lb, disk and cross-disk to distribute fertilizer and aerate treatment zone.
8. Irrigate as necessary to maintain treatment zone moisture between 50 and 80% of field capacity.
9. After 6 to 8 weeks sample site to demonstrate cleanup level <1% TPH. Retreat as above if necessary, based on after remediation sampling and analyses.

8.6   SCENARIO II—HEAVY CRUDE OIL SPILL ON SOIL

Spill onto native soil, 20 barrels of heavy crude oil, gravity 10° API.

8.6.1   Considerations

Heavier crudes are not as amenable to biodegradation as the lighter weight crudes. To compensate for this probem, reduce the initial petroleum hydrocarbon loading rate for 10–20° API crude to 3% TPH, based on 22°C soil temperature. The temperature considerations for TPH loading rates are shown in Table 8.2.

Microbial degradation of oil does not occur below 5°C soil temperature. However, composting is possible. Add an equivalent amount of organic matter, e.g. peat moss or cellulose fiber, as an extender and adsorbent. The added carbon sequesters the more labile TPH fractions and added nutrients. This provides a more favorable soil environment for colonization of indigenous microflora and proliferation of oil "eaters". Animal manure may also be used. Fresh manures require more tillage operations to maintain an aerobic system, since animal manures contain low C:N ratios and degrade rapidly.

8.6.2   Site Assessment

Crude oil with an API gravity of 10° has a specific gravity of 1. A 20 bbl spill having an average concentration of 10% oil to a depth of 4 in. will occupy an area approximately 2294 ft^2 (48 ft × 48 ft). Mix the contaminated soil with background soil to achieve the recommended 3% loading rate (Table 8.3). The following procedure demonstrates one way of controlling the application rate, while limiting the treatment zone to the biologically active surface. A treatment zone depth of 5 in. is selected for calculation and design purposes. This depth of incorporation and mixing is readily achieved using conventional farm equipment.

The proper loading rate is ahieved by a combination of removal and fill operations, which result in a total treatment area (inclusive of the original spill boundary) of 6098 ft^2 (127 ft × 48 ft).

8.6.3   Calculations

1. Assumptions:

- Soil temperature > 22° C, TPH loading rate = 3%
- Acre-6 in. = 2,000,000 lb (91.83 lb/ft^3)
- Oil equally distributed in 4-in. profile at 10%

TABLE 8.2
INITIAL TPH LOADING RATES FOR OIL;
<20° API

Mean Annual Soil Temperature, °C	Loading Rate, % TPH
≥22.0	3
15.0–21.9	2
8–14.9	1
<8	0

TABLE 8.3
PROPERTIES OF PRODUCED
WATER FROM WEST TEXAS

Parameter	Value
pH, s.u.	6.8
EC, mmhos/cm	251
Na, mg/L	42,869
Ca, mg/L	13,848
Mg, mg/L	2437
Cl, mg/L	88,147
SO_4, mg/L	991
HCO_3, mg/L	327

- Background soil oil residual 0.01%
- 5-in. depth of mixing
- Crude oil contains 78% carbon

2. Contaminated soil mass

$$20 \text{ bbl} \times 42 \text{ gal/bbl} \times 3.81 \text{ L/gal} = 3200 \text{ L}$$
$$3200 \text{ L} \times 1 \text{ kg/L} \times 2.2 \text{ lb/kg} = 7041 \text{ lb oil}$$
$$(7041 \text{ lb oil}/\times \text{ lb soil}) \times 100 = 10\% \text{ oil in soil}$$
$$X = 70,410 \text{ lb contaminated soil}$$

3. Background soil requirement for controlled loading to 3%

$$\text{TPH, 3\%} = \frac{[(10\%)(70,410 \text{ lb}) + (0.01\%)(X \text{ lb soil})]}{(70,410 \text{ lb} + X \text{ lb soil})}$$
$$211,230 \text{ lb} + 3X = 704,100 + 0.01X$$
$$X = 169,955 \text{ lb background soil}$$
$$\text{Treatment area mass} = 169,955 \text{ lb} + 70,410 \text{ lb}$$
$$= 240,365 \text{ lb}$$
$$\text{Area equivalent} = 240,365 \text{ lb}/1,666,667 \text{ lb/acre-5 in.}$$
$$= 0.1442 \text{ acre} \cong 0.14 \text{ acre}$$
$$= 0.14 \text{ acre} \times 43,560 \text{ ft}^2/\text{acre}$$
$$= 6098 \text{ ft}^2 (48 \text{ ft} \times 127 \text{ ft}) \text{ assuming spill area is 48 ft wide}$$

4. Carbon content in ppm (mg/kg)

$$\text{C\%} = \text{TPH\%} \times 0.78$$
$$\text{C\%} = 3.0\% \times 0.78$$
$$\text{C\%} = 2.3\%$$
$$\text{C mg/kg} = 2.3\% \times 10,000$$
$$\text{C mg/kg} = 23,000 \text{ mg/kg}$$

5. Nitrogen requirement

$$\text{C:N} = 150:1$$
$$(23,000 \text{ mg/kg})/N = 150:1$$
$$N = 153 \text{ mg/kg}$$
to convert from mg/kg to lb/acre-6 in multiply by 2
$$N = 153 \text{ mg/kg} \times 2$$
$$N = 306 \text{ lb/acre-6 in.}$$
$$N = 255 \text{ lb/acre-5 in.}$$

6. Treatment (spill + receiving) area occupies 0.14 acre (Figure 8.1).

$$N = 255 \text{ lb/acre-5 in.} \times 0.14 \text{ acre}$$
$$N = 36 \text{ lb}$$

7. NPK 4:1:1

$$N/4 = P,K = 36/4$$
$$P,K = 9 \text{ lb}$$

8.6.4   Procedure (See Figure 8.1)

1. Remove all pooled oil from spill area (Step 1).
2. Excavate 1.5 in. soil from receiving area adjacent to spill and place in stockpile area (Figure 8.1 Step 2),
3. Remove 2.5 in. from spill area to receiving (treatment) area. Spread and level to grade (Figure 8.1, Step 3),
4. Backfill spill area with uncontaminated soil stockpiled in holding area. Spread and level to grade (Figure 8.1, Step 4).
5. Add half of the ammonium nitrate, 55 lb (33% N), all of the super phosphate, 98 lb (9.2% P) and all of the muriate of potash, 18 lb (51%) to the spill and receiving areas.
6. Disk and cross-disk to 6 in. to distribute fertilizer and aerate treatment zone.
7. After 6–8 weeks add half of the remaining ammonium nitrate (27 lb), disk and cross-disk to distribute fertilizer and aerate treatment zone.

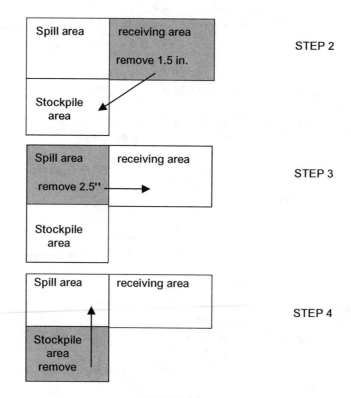

FIGURE 8.1
SOIL MOVING PLAN FOR SCENARIO II

8. After 6 to 8 weeks add final 27 lb ammonium nitrate, disk and cross-disk to distribute fertilizer and aerate treatment zone.
9. Irrigate as necessary to maintain treatment zone moisture between 50 and 80% of field capacity.
10. After 6–8 weeks sample site to demonstrate cleanup level <1% TPH. If the 1% TPH is not achieved, retreat treatment as above.

8.7 SCENARIO III—LIGHT CRUDE OIL AND PRODUCED WATER SPILL ON SOIL

West Texas crude oil plus produced water were spilled onto native soil. The mixture is 40 barrels of water with 10 barrels of oil.

8.7.1 Considerations

Oil distribution from the spill event involving mixture of oil and water is controlled by the amount of water, infiltration capacity and soils' capacity to transmit water through the profile. Typically, infiltration capacities are higher than percolation rates. This creates a condition of surface saturation and causes spreading of contamination during spill events.

Produced waters typically are high in sodium chloride (NaCl). Typical produced waters from West Texas oil fields have following properties (Table 8.3).

The following equation approximates specific conductance (SPEC) and total salt in solution (total dissolved solids, TDS):

$$TDS = SPEC \times 613$$

Where 613 is a constant determined empirically for oil field produced waters. The summation of ions (Table 8.3) yields a TDS of 148,619 mg/L, comparing well to the calculated value of 153,863 mg/L (251 × 613). Forty barrels produced water at 153,863 mg/L TDS is equivalent to 2161 lb salt [153,863 mg/L × 3.8 L/gal × 42 gal/bbl × 2.2 lb/kg × kg/10^6 mg × 40 bbl]. The impact of this much salt spilled on the land is profoundly influenced by the area, and eventually the matrix volume of the receiving soil.

Electrical conductivity evaluated from the spill area soil saturated paste provides the best measure of salt impact. Saturated paste EC (SPEC) values for the spill area and receiving soil texture are calculated based on the affected soil receiving 0.06, 0.6 and 1.2 in. brine/acre, percolated to a depth of 6 in. (Table 8.4).

TABLE 8.4
EFFECTS OF SALT AND TEXTURE ON SATURATED PASTE
ELECTRICAL CONDUCTIVITY

Total Salt, lb	Spill Area, acre	Textural Class	SP Moisture, %	SPEC, mmhos/cm
2161	0.05	Sand	28	127
2161	0.05	Loam	38	93
2161	0.05	Clay	48	74
2161	0.10	Sand	28	63
2161	0.10	Loam	38	47
2161	0.10	Clay	48	37
2161	1.00	Sand	28	6.3
2161	1.00	Loam	38	4.7
2161	1.00	Clay	48	3.7

On-site treatment options for restoring the land to full agronomic potential include rinse and fill, in situ leach, dilution burial and land spreading techniques.

Oil floating on the water is assumed to affect the same spatial volume as the produced water. The loading is approximately 3.4% oil in soil for a 10-barrel oil spill onto 0.05 acres. Corresponding oil loadings for 0.1 acre is 1.7 %, and 1 acre is 0.2%. TPH levels of 3.4% and 1.7% require treatment to 1% within 1 year under Railroad Commission of Texas (RRC) Statewide Rule 91.

8.7.2   Fertilizer

Nitrogen, phosphorus and potassium fertilizers are purchased in combination with other elements. The most common combinations (balanced fertilizers) are groupings of the three, for example, 10-10-10 or 12-12-12. These numbers are read as 10% N, 10% P_2O_5, 10% K_2O. P_2O_5 contains 44% phosphorus, and K_2O contains 83% potassium. Thus, 10-10-10 represents 0.10 N, 0.044 (0.10 × 0.44) P and 0.083 (0.10 × 0.83) K. Similarly, superphosphate contains 9.2% (0.092) P and ammonium nitrate 33% N. These percentages come directly from the atomic weights of the elements.

8.7.3   Remediation Procedure

8.7.3.1   Minimum Management Method.   SPEC values between 8 and 16 mmhos/cm significantly reduce agronomic use of the land, Total loss occurs when SPEC >30 mmhos/cm. Given the small acreage involved in this scenario, it may be more prudent to abandon the land as wasted and plant nonagronomic halophytic (salt tolerant) shrub seedlings (e.g., fourwing saltbush, Atriplex canescens; winterfat, Ceratoides lanata; prostrate kochia, Kochia prostrata) to stabilize the landscape. This is considered a minimum management and is not intended to restore agronomic potential.

8.7.4   Minimum Management Method

8.7.4.1   Calculations.   Fertilizer calculations and application rates for minimum management procedure:

1. Carbon content in ppm (mg/kg)

$$C\% = TPH,\% \times 0.78$$
$$C\% = 3.4\% \times 0.78$$
$$C\% = 2.65\%$$
$$C\ mg/kg = 2.65\% \times 10,000$$
$$C\ mg/kg = 26,500\ mg/kg$$

2. Nitrogen requirement

$$C:N = 150:1$$
$$(26,500\ mg/kg)/N = 150:1$$
$$N = 177\ mg/kg$$
$$N = 177\ mg/kg \times 2$$
$$N = 354\ lb/acre\text{-}6\ in.$$

3. Treatment area occupies 0.05 acre

$$N = 354\ lb/acre\text{-}6\ in. \times 0.05\ acre$$
$$N = 18\ lb$$

4. N:P:K 4:1:1

$$N/4 = P,K = 18\ lb/4$$
$$P,K = 5\ lb$$

To obtain the necessary K and P concentrations use 60 lb of 10-10-10 fertilizer. This contribution provides 6.0 lb of N (60×0.10); 2.6 lb. of P (6.0×0.44); and 5 lb. of K (6.0×0.83). Supplement the 2.6 lb of P with 26 lb. (2.4/0.092) of superphosphate ($Ca(H_2PO_4)_2$) For the first application add all of the P and K and half of the total N (18 lb /2). The 10-10-10 fertilizer adds 6 lb of N so only 3 lb of N need be added as ammonium nitrate (33% N).

Ammonium nitrate = (3 lb/0.33) = 9 lb

8.7.5   Procedure

1. Apply 2.5 tons good-quality hay or cellulose fiber to spill area.
2. Apply 9 lb ammonium nitrate, 60 lb of 10-10-10 balanced fertilizer and 26 lb of superphosphate to spill area.
3. Disk and cross-disk to 6 in. to distribute organic matter and fertilizer and aerate the soil.
4. After a 6- to 8-week interval, apply 14 lb ammonium nitrate (33% N) and disk to distribute fertilizer and aerate soil.
5. After a 6- to 8-week interval, apply 14 lb ammonium nitrate and disk to distribute fertilizer and aerate soil.
6. Irrigate as necessary to maintain soil between 50 and 80% of field capacity.
7. After 6–8 weeks from last fertilizer application sample and analyze to demonstrate TPH <1%. Repeat treatment, if necessary.
8. Plant seedling halophytes on 2-m (6.6-ft) centers

8.7.6   Rinse and Fill Method

This technique is used for low-volume, high-brine, coarse or readily flocculated soils. Soils are excavated, suspended in a high-calcium solution (0.1N calcium chloride, e.g., 3.6 lb $CaCl_2$/bbl) and flocculated with a centrifuge operation. Treated solids, having ECs between 8 and 16 mmhos/cm, are returned to the excavation and mixed with an equal volume of undisturbed soil to complete salt amelioration. Solids with EC >16 are reflocculated. Hydrocarbon is biodegraded to complete site restoration, if not removed during the rinse operation.

8.7.7   Procedure

1. Remove salt-contaminated soil (40 cu yd from the 0.05 acre spill area) to a large mixing container.
2. Blend soil with 0.1N $CaCl_2$ solution to suspend solids and displace brine.
3. Centrifuge to separate solids from rinsate.
4. Separate oil from rinsate and rinsate by well injection. Return the oil to the hydration stream.
5. Return solids having <16 mmhos/cm EC to excavated area, blending with an equal volume of native soil. Blend by disking to 6 in.
6. Add 200 lb high-quality hay or cellulose fiber to increase aeration and organic matter. Add 50 lb 10-10-10 balanced fertilizer and disk to 6 in. Sample and analyze for petroleum hydrocarbon to verify TPH <1%. Repeat if necessary.
7. Seed site with native grasses.

8.7.8   *In Situ* Leach Method

In situ leach is applicable when a spill occurs on deep well-drained soils, with low-salt leachwater available and no threat exists to usable groundwater. Remove salts from shallow soils or soils with an impeding layer by directing artificial drainage to a collection sump for disposal. Also, artificial drainage can be used to protect groundwater resources by intercepting brackish leachate. The soil profile from the surface downward must be sufficiently permeable for drainage to occur.

The spacing between drain lines will be determined by the hydraulic conductivity and depth to restrictive layer or desired rooting profile. As a general rule, spacing of laterals ranges from 15 to 60 ft. One and one-half pore volumes of water are required to remove excess salt from the profile. The amount of water to be drained and the available time are also important design considerations. Installing drains 2–3 ft deep increases the number of laterals, but decreases the volume of water necessary to flush salt. Do not bury drains more than twice the depth of the surface layer in stratified soil (e.g., coarse-textured surface over finer-textured subsurface layer).

Gypsum application rates can be determined either from sodium adsorption ratio (SAR) as it relates to the exchangeable sodium percentage (ESP), or it may be calculated from actual ESP measurements. The latter is usually employed under nonequilibrium conditions, for example, soil recently impacted.

8.7.9   Calculation of SAR

Calculation of SAR is from the analysis of produced water (Table 8.3). Convert to mEq/L from mg/L:mEq/L is equal to mg/L/mEq wt (See Appendix I, Annex G).

8.7.10   Calculations

Sodium, Na^+ 42,869 mg/L/23 mg/mEq = 1,864 mEq/L
Calcium, Ca^{2+} 13,848 mg/L/20 mg/mEq = 692 mEq/L
Magnesium, Mg^{2+} 2,437 mg/L/12 mg/mEq = 203 meEq/L
SAR = Soluble Na (mEq/L)/[soluble Ca (mEq/L) + soluble Mg (mEq/L)/2]$^{1/2}$
SAR = 1,864/[(692 + 203)/2]$^{1/2}$ = 88

8.7.11   Calculation of ESP from SAR

$$ESP = [100(-0.0126 + 0.01475 \times SAR)]/[1 + (-0.0126 + 0.01475 \times SAR)]$$

$$ESP = [100(-0.0126 + 0.01475 \times 88)]/[1 + (-0.0126 + 0.01475 \times 88)] = 56\%$$

This calculation indicates a soil in contact with oilfield produced water having an SAR of 88 will equilibrate to an ESP of 56%.

TABLE 8.5
GYPSUM REQUIREMENTS FOR VARIOUS SOIL TYPES

Texture	CEC, mEq/100 g	SAR	ESP, %	Gypsum Requirement lb/0.05 acre-ft
Sand	3.5	88	56	262
Loam	15.0	88	56	1122
Clay	35.0	88	56	2618

8.7.12   Gypsum requirement determined from CEC and ESP

It is undesirable for the ESP to exceed 12% of the CEC. To compensate, Ca^{2+} is added in the form of gypsum at a rate equivalent to the actual exchangeable sodium minus the desired exchangeable sodium. Gypsum requirement will vary dependent on soil type and corresponding CEC (Table 8.5).

$$\frac{(\text{Actual \% ESP} - \text{Desired \% ESP})}{100} \times \text{CEC} = \text{CR (charge requirement)}$$

$$\frac{(\text{Actual \% ESP} - 12\%)}{100} \times \text{CEC} = \text{CR (charge requirement) (Tons/acre-ft)}$$

Divide the charge requirement by 2 for determining tons per acre-6 in.
This following example is based on data for sand (Table 8.5).

$$\frac{(56\% \text{ ESP} - 12\% \text{ ESP})}{100} \times (3.5 \text{ mEq/100 g CEC}) = 1.54 \text{ (charge requirement)}$$

$$\text{Gypsum requirement} = 1.54 \times 1.7 \text{ (tons } CaSO_4\text{/acre-ft)}$$
$$= 2.6 \text{ (tons } CaSO_4\text{/acre-ft)}$$

8.7.13   Procedure

1. Install drain lines 2 to 3 ft deep at spacing designed to remove the desired quantity of water within a given time interval.
2. Laterals are drained to a common sump line draining. Drain lines are covered with rock or gravel followed by soil and contoured to grade.
3. Add gypsum at the application rates calculated (Table 8.5)
4. Apply 50 lb of 10-10-10 balanced fertilizer and disk to distribute and aerate.
5. Wet soil with water and maintain between 50 and 80% of field capacity during biodegradation phase, for example, 1 in. per week.
6. After 6–8 weeks apply 15 lb ammonium nitrate and disk to distribute and aerate.
7. After 6–8 weeks apply 15 lb. ammonium nitrate and disk to distribute and aerate.
8. Chisel soil to 16 in. following a 6- to 8-week reaction interval.
9. Impound salt affected soil with a levee and flood irrigate with 4 in. of fresh water (5430 gal/0.05 acres).
10. After water penetrates surface, flood again with 4 in. to effect salt removal.
11. Commingle leachate collected in sump with oil field produced water for well injection.

8.7.14   Dilution Burial Method

Dilution burial applied to salt and hydrocarbon contaminated soil is designed to reduce salinity to <24 mmhos/cm SPEC and hydrocarbon levels to <1% TPH. Groundwater depth should be at least 40 ft below land surface for SPEC >24 mmhos/cm, and 20 ft for an SPEC ≤12 mmhos/cm. The bottom of the burial trench should be at least 25 ft above groundwater for SPEC >24 mmhos/cm and 5 ft for an SPEC ≤12 mmhos/cm. The top of the impacted/clean soil mix should be 5 ft below the land surface.

8.7.15   Assumptions and Calculations

Spill area = 0.05 acres
Depth of penetration = 6 in.
Depth to groundwater > 40 ft

8.7.16   Calculations

VALUES TO USE IN CALCULATIONS

Soil Value	EC, mmhos/cm	ESP, %	TPH, %
Spill area	127	25.0	3.2
Background	0.8	1.0	<0.01
Target threshold	24	12	1

8.7.17   Calculations: Limiting Constituent

1. *Contaminated soil, volume*

$$\text{Soil volume} = 0.05 \text{ acres} \times 43{,}560 \text{ ft}^2/\text{acres} \times 0.5 \text{ ft}$$
$$= 1089 \text{ cu ft}$$

2. *Soil volume requirement* for salinity management using mass balance equation

$$24 \text{ EC} = (127 \text{ EC})(1089 \text{ cu ft}) + (X \text{ cu ft})(0.8 \text{ EC})/(1089 \text{ cu ft} + X \text{ cu ft})$$
$$X = 4835 \text{ cu ft}$$

3. *Soil volume requirement* for ESP Management

$$12 \text{ ESP} = (25 \text{ ESP})(1089 \text{ cu ft}) + (X \text{ cu ft})(1.0 \text{ ESP})/(1089 \text{ cu ft} + X \text{ cu ft})$$
$$X = 4026 \text{ ft}$$

4. *Soil volume requirement* for organic management

$$1 \text{ TPH} = \frac{(3.2\,TPH)(1089ft^3) + (Xft^3)(0.01\,TPH)}{(1089ft^3 + Xft^3)}$$
$$X = 2420 \text{ cu ft}$$

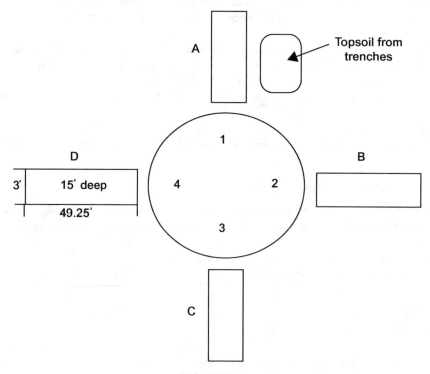

FIGURE 8.2
TRENCH CONSTRUCTION SCHEMATIC

Salinity, as shown by the above calculations, is the limiting constituent (largest dilution volume). The dilution requirement is 4,835 cu ft. This plus the contaminated soil volume of 1,089 cu ft yields a total trench volume of 5,924 cu ft. A 15-ft trench depth is required for 10 ft of storage depth and 5 ft for cover. A trench 3 ft wide × 197 ft long will accommodate 5910 cu ft of impacted/clean soil mix. Installing several radial trenches, totaling 197 ft, in an array surrounding the spill area expedites reclamation. For example, we assumed the 0.05-acre spill is circular and four trenches (49.25 ft each) are opened to receive the impacted soil/native soil mix (Figure 8.2). No treatment for ESP is required. However, we recommend a blanket application of gypsum be made based upon soil type (Table 8.6).

8.7.18   Procedure

1. Open trench A Figure 8.2), 49.25 ft long × 15 ft deep × 3 ft wide, segregating the initial 6.8 ft of material removed for use as backfill, topsoil cover.
2. Excavate contaminated soil area 1 (1089/4 = 272.25 cu ft) and place in trench A blending mechanically with 1208.75 cu ft (4835/4) of uncontaminated mix soil.
3. Repeat steps 1 and 2 for trenches B, C and D and excavation soil areas 2, 3 and 4, respectively.
4. Backfill trench areas with cover material, placing the deeper excavated materials in first, followed in sequence with topsoil replaced last.
5. Remaining topsoil is used as backfill in the spill area.
6. Apply 50 lb/acre 13-13-13 balanced fertilizer to the disturbed area. Area may be seeded or allowed to revegetate naturally.

8.7.19   Land Spreading Method

Land spreading is a technique for rendering nonhazardous waste solids and contaminated soils harmless through soil incorporation. This method uses dilution, chemical alteration, physical adsorption and biodegradation processes in reducing constituents to acceptable levels consistent with intended land use.

Method 1

Assumptions: Method No. 1:

$$\text{spill area} = 0.05 \text{ acres}$$
$$\text{Depth of penetration} = 6 \text{ in.}$$
$$\text{Depth to groundwater} > 40 \text{ ft}$$

TARGET VALUES TO USE IN CALCULATIONS

Soil Value	EC, mmhos/cm	TPH, %
Spill area	127	3.2
Background	0.8	<0.01
Target Threshold	8	1

a. Calculation No. 1: limiting constituent

1. Contaminated soil, volume

$$\text{Soil volume} = 0.05 \text{ acres} \times 43,560 \text{ ft}^2/\text{acres} \times 0.5 \text{ ft}$$
$$= 1089 \text{ cu ft}$$

2. *Soil volume requirement* for salinity management

$$8 \text{ EC} = (127 \text{ EC})(1089 \text{ cu ft}) + (X \text{ cu ft})(0.8 \text{ EC})/(1089 \text{ cu ft} + X \text{ cu ft})$$
$$X = 17,999 \text{ cu ft}$$

TABLE 8.6
BLANKET GYPSUM APPLICATIONS
FOR VARIOUS SOIL TYPES

Textural Class	Gypsum Requirement, lb/acre
Sand	150
Loam	500
Clay	1500

3. *Soil volume requirement* for organic management

$$1 \text{ TPH} = (3.2 \text{ TPH})(1089 \text{ cu ft}) + (X \text{ cu ft})(0.01 \text{ TPH})/(1089 \text{ cu ft} + X \text{ cu ft})$$
$$X = 2420 \text{ cu ft}$$

Salinity, as shown by the above calculations, is the limiting constituent. The dilution requirement is 17,999 cu ft. To incorporate to 6 in. requires 0.88 acres of land. 0.88 acre = (19,087 cu ft)/(21,780 cu ft/acre 6-in)

b. Procedure No. 1

1. Scrape contaminated soil from the spill area and spread evenly over 0.88 acres to a depth of 0.3 in. Strongly structured and massive soils will be difficult to spread thinly and may require a thicker initial application on less acreage followed by redistribution over 0.88 acres.
2. Disk the treatment area to a depth of 6 in.
3. Apply a blanket application of 136 lb gypsum ($CaSO_4$) to the treatment area and disk again.
4. The excavated spill area is restored to the original contour with treated soil.
5. Apply 75 lb ammonium nitrate and 50 lb of 13-13-13 fertilizer to the disturbed area. Area may be seeded or allowed to revegetate naturally.

Method No. 2
Assumptions: Method No. 2:

$$\text{Spill area} = 0.1 \text{ acres}$$
$$\text{Depth of penetration} = 6 \text{ in.}$$
$$\text{Depth to groundwater} > 40 \text{ ft}$$

VALUES TO USE IN CALCULATIONS

Soil Value	EC, mmhos/cm	TPH, %
Spill area	63	1.7
Background	0.8	<0.01
Target threshold	8	1

a. Calculations No 2. Limiting constituent

1. *Contaminated soil, volume*

$$\text{Soil volume} = 0.1 \text{ acres} \times 43,560 \text{ ft}^2/\text{acres} \times 0.5 \text{ ft}$$
$$= 2,178 \text{ cu ft}$$

2. *Soil volume requirement* for salinity management

$$8 \text{ EC} = (63 \text{ EC})(2178 \text{ cu ft}) + (X \text{ cu ft})(0.8 \text{ EC})/(2178 \text{ cu ft} + X \text{ cu ft})$$
$$X = 16,638 \text{ cu ft}$$

3. *Soil volume requirement* for organic management

$$1 \text{ TPH} = (1.7 \text{ TPH})(2178 \text{ cu ft}) + (X \text{ cu ft})(0.01 \text{ TPH}) / (2178 \text{ cu ft} + X \text{ cu ft})$$

$$X = 1{,}540 \text{ cu ft}$$

Salinity, as shown in the above calculations, is the limiting constituent. The dilution requirement is 16,638 cu ft. To incorporate to 6 in. requires 0.76 acres of land.

b. Procedure Method No. 2

1. Scrape contaminated soil from the spill area and spread evenly over 0.76 acres to a depth of 0.40 in. Strongly structured and massive soils will be difficult to spread thinly and may require a thicker initial application on less acreage followed by redistribution over 0.76 acres.
2. Disk the treatment area to a depth of 6 in.
3. Apply a blanket application of 136 lb gypsum ($CaSO_4$) to the treatment area and disk again.
4. Restored the excavated spill area is to the original contour with treated soil.
5. Apply 75 lb ammonium nitrate and 50 lb of 13-13-13 fertilizer to the disturbed area. Area may be seeded or allowed to revegetate naturally.

8.7.20 Vegetative Restoration

8.7.19.1 Objective. The final step to site closure is establishing a vegetative cover. A vegetative cover serves two purposes: (1) erosion control and (2) aesthetically pleasing landscape. The latter is important since it instills a degree of confidence that no significant levels of contamination linger onsite.

8.7.21 Tillage Requirements

Grass seed are small and some effort is required to establish an adequate seedbed. Disk and harrow the seedbed to fracture the soil into sufficiently small aggregates so as to provide good seed and soil contact. Primarily, the moisture level controls tillage of the soil. If the soil is worked wet, particularly when mixing already dispersed materials, the soil becomes puddled resulting in a near structureless condition. If worked too dry, clods may be formed that are not readily fractured.

TABLE 8.7
WEST TEXAS PLANT SPECIES SEEDING RATES

Season	Plant Species	Seeding Rate, lb/acre
Spring/summer	Common bermudagrass	10
	Kleingrass-75	20
	Weeping lovegrass	5
	Clare subterranean clover	8
Fall/winter	Common bermudagrass	10
	Kleingrass-75	20
	Weeping lovegrass	5
	Annual ryegrass	40
	Clare subterranean clover	8

8.7.22   Seeding

Grass seed should be broadcast on roughened surfaces and rolled. Germination and seedling survival is improved with a hay mulch cover. Do not work seed into soil deeply. The planting depth should not exceed 1/2 in. and is best if it can be controlled between 1/8 and 1/4 in.

8.7.23   Fertility Requirements

The seeding formulations given in Table 8.7 are designed for quick ground cover and a degree of permanence. Permanence is controlled ultimately by the fertility status of the growth media. Typically, native soils have a low innate fertility relative to nitrogen, phosphorus and potassium. Accordingly fertilizer is required to ensure adequate fertility.

We recommend applying a minimum 300 lb/acre of 10-10-10 or 12-12-12 balanced fertilizer during the tillage operation. Adding clover to the seed formulations aids maintenance of nitrogen fertility. However, a top-dress application of 30 lb of nitrogen may be required 6–8 weeks following emergence to ensure the clover survives.

REFERENCE

EPA. 1987. "Technical Report, Exploration, Development, and Production of Crude Oil and Natural Gas." EPA 530-SW-87-005. Office of Water Regulations and Standards (WH-552), Washington, DC.

CLEAN CLOSURE TECHNIQUES

9.1 INTRODUCTION

With the increased environmental and regulatory sensitivity associated with the operation of earthen pits, it would be proactive for an operator working in a state with minimum regulations to adopt clean closure practices that embrace the best engineering and cost-effective techniques available. Obviously the goal is to close pits using responsible and environmentally sound procedures in accordance with applicable laws, regulations and/or agreements with the land owner(s).

Closure techniques addressed in this chapter are intended to provide oil and gas operators and Regulatory Agency with information on the key steps in designing and implementing pit closures. These guidelines are intended for drilling reserve pits, but the procedure apply to emergency pits, and the decommissioning and closure of production pits. On-site waste management techniques and limiting constituent guideline values presented in this chapter and other chapters are applicable to E&P wastes in general and contaminated soils associated with leaks and spills of saltwater, petroleum hydrocarbon and reserve pit liquids.

9.2 SITE CLOSURE CONSIDERATIONS

When making the decision to close a drilling reserve, emergency, and production pit; the operator must give careful consideration to selecting the closure method and waste disposal options, which are most practical from a personnel, materials, equipment, and cost perspective and minimizes the environmental impact. Factors to consider are:

- Inventory of waste constituents
- Treatment/disposal options available
- Depth to ground water
- Landowner considerations including post closure land use
- Ministry or regulatory approval
- Minimizing environmental impact
- Cost

Recommendations in this book are *conservative by design*. Selected disposal and closure methods are based on the Guidelines for Limiting Constituent (GLC) values (conceptually presented in Chapter 7) developed for the various land resources impacted with no limitation for future land use.

9.3   DECOMMISSIONING

When pit usage is discontinued decommissioning should proceed with the thought the pit may be used as a fresh water impoundment for recreational purposes or for watering livestock. If these uses are not a consideration, decommissioning should proceed as follows:

- Identify and stop all oilfield waste discharges to the pit
- Remove free oil and/or water for appropriate disposal
- Remove solid debris (concrete, wood, etc.) interfering with closure
- Remove waste solids exceeding GLC values or treat solids *prior* to closure
- Mix pit waste with pit wall soil to <50% moisture content
- Backfill and restore original grade and contour
- Remove all signs associated with pit use

9.3.1   Sampling Pit Liquids

Pit liquids are sampled after recovery of free oil. The purpose is to test the aqueous phase to determine its quality for use as a water resource. Pit liquids are sampled at multiple points (Chapter 4), blended together and the resultant composite analyzed in the field for electrical conductivity (EC). Liquids having an EC <2.4 mmhos/cm (1500 mg/L total dissolved solids) are suitable as a one-time irrigation source for use in production agriculture or for enhanced bioremediation of oily pit solids. Liquids having an EC > 2.4 mmhos/cm can be land applied on a one-time basis using GLC analysis and management. Slight or moderately saline wastewaters should be commingled with production brine for disposal in an injection well. *__Do not__* inject low-salt pit water without mixing with produced water or other brine fluid first.

9.3.2   Sampling Pit Solids

Pit waste solids offer more of a challenge than sampling pit liquids. The challenge is to collect representative samples; since disposal options and treatments are based on the assumption the samples represent the pit materials. Poor or inadequate sampling can lead to erroneous conclusions regarding the disposal and/or treatment options. Improper disposal could lead to additional clean-up efforts and costs. It is important to recognize that constituent levels are about 20–30 times greater in pit solids than in pit liquids. Thus, as discussed in Chapter 5, sampling design is particularly important in large pits.

If the pit solids are soft and cannot support a person's weight, access from boat or pad to collect samples is indicated. It is good practice to access a large pit having surface water from a boat after the free oil has been removed. As discussed in Chapter 5, samples can be obtained by pushing a hollow tube of aluminum conduit or polyvinyl chloride (PVC) pipe, open at both ends, through unconsolidated pit solids to the consolidated native soil forming the pit floor. The tube needs to be of sufficient length to sample through the aqueous phase and through the unconsolidated pit solids and into native soil. The tube is pushed into the native soil to create a plug for retrieval of the cross section of pit liquids and pit solids. These are discharged from the tube to a common container where the liquid phase can be decanted back into the pit. Do not combine the soil plug with the pit solids sample. Save the native soil plugs in a separate container.

Pit solids are combined to form discreet composite samples. The larger the pit the greater will be the need for composites to extract information regarding materials disposal and treatment. Usually one sample composite representative of the entire pit will suffice for hazardous waste characterization.

Sample composites representative of smaller unit areas (~30 m^2 or 325 ft^2) within a large pit are analyzed individually for index parameters necessary for waste management. Saturated paste EC, ESP and total petroleum hydrocarbon (TPH) are the most frequently used index parameters.

If the pit solids are firm and will support a person's weight, a small 5-cm diameter stainless steel auger may be used to obtain the necessary samples. The entire depth of waste solids should be collected at each sample point within the pit. This usually requires multiple trips to the same sample hole. The under lying native soil should be sampled to a recommended depth of 0.5 m or at least 1.5 times the depth of the pit solids, whichever is greater.

9.3.3 Analyses

State agencies, typically, require pit solids be analyzed for one or more of the following parameters:

- Moisture
- Ignitability
- Corrosivity (pH)
- Reactive sulfides
- Total benzene, toluene, ethyl benzene and xylenes (BTEX)
- TCLP BTEX
- Oil & grease (O&G)
- Total petroleum hydrocarbon (TPH)
- Chlorides
- Salinity
- Total metals (As, Ba, Cd, Cr, Cu, Hg, Ni, Ag, Pb, Se, Zn)
- Toxic characteristic leaching procedure (TLCP) metals

The most important of the above parameters are TPH and salinity, since the affect plant growth.

In addition, we recommend pit solids be analyzed for the saturated paste moisture equivalent. After reaction with the solids the paste liquid is extracted and analyzed for SPEC, and soluble cations calcium, magnesium and sodium, which are used to calculate the sodium adsorption ratio (SAR). If the SAR is >15, we recommend the cation exchange capacity (CEC) be determined for the pit solids. SAR and CEC are used to calculate calcium amendments necessary for treatment and remediation of saline-sodic pit solids. For our management goals, saline materials are defined as pit solids yielding a saturated paste EC >4 mmhos/cm. Sodic materials are defined as pit solids having an SAR >15. *Experience shows the parameters potentially limiting E&P waste disposal typically are salinity (EC), sodicity (ESP), TPH and total metals barium, chromium, lead and zinc.*

9.4 CLOSURE OPTIONS

We list and describe as many pit closure options as possible. This is an area of active research and field demonstration as well as ongoing review by state agencies charged with regulating earthen pit utilization and pit closure.

9.4.1 Pit Liquids Disposal

Inspect the pit to determine if liquids are present. This includes determining liquid phase(s) volume in the pit. Skim the free oil and removed to the production stream. Sample and analyze the water phase to determine if it is fresh enough to be used in the field as an irrigation source for pit solids remediation. Bioremediating hydrocarbon impacted solids requires fresh water to support biological degradation. When fresh water is in limited supply during the dry season, the nonsaline pit water is a potential irrigation source. If the pit water is too salty remove and commingle it with produced water for disposal in an injection well. You need to document pit liquid volumes, salinity (EC), and method of use or disposal, e.g., injected or recycled, Figure 9.1, to establish the liquid were handled correctly.

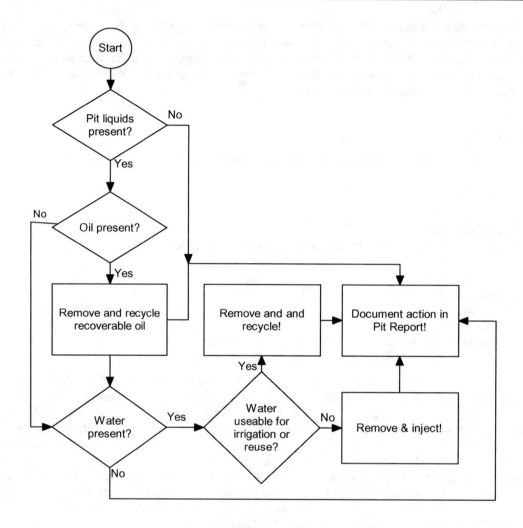

FIGURE 9.1
PIT LIQUID DISPOSAL FLOWCHART

9.4.2   Pit Solids Disposal

EPA (1980) and API sponsored studies of drilling reserve pits; emergency pits and production pits showed the bulk of these materials are nonhazardous and are suitable for on-site pit closure techniques. Pit closure/solids disposal options are:

- Minimum management
- Burial
- Dilution burial
- Land treatment (also known as land farming or land spreading)
- In situ leach for saline pit solids
- In situ bioremediation for oily waste solids
- Solidification/chemical stabilization
- Road spreading
- Third-party disposal

9.5 MINIMUM MANAGEMENT

Minimum management is a process reserved for pit solids at or below recommended GLC values (Table 9.1). The basic technology entails blending pit solids with berm materials, after removing any free liquids. Asphaltic layers are scrapped from the liquid surface and used as base material in facility construction or lease road repair. Before mixing waste solids with berm materials the pit surface is dusted with agricultural gypsum ($CaSO_4 \cdot 2H_2O$) at a rate of 540 kg/ha (500 lb/acre) or agricultural limestone ($CaCO_3$) as needed to raise the pH to 5.5–6 s.u. Additional topsoil is brought-in, spread and contoured as necessary to restore the site to the original surface grade and contour. Backfilling is followed by a nominal application of balanced fertilizer (12-12-12 or 13-13-13) disked into the contoured surface to encourage revegetation. A mulch application followed by seeding or sprigging with native grasses may be necessary to stabilize the surface and prevent erosion.

9.5.1 No-Closure Option

No-closure option is a decision made under the minimum management GLC guidelines, Table 9.1, allowing an *uplands* pit to remain open for recreational or alternative use benefit (farming) as a fresh water pond. Pit solids at or below these criteria (Table 9.1) pose no limitation on plant growth, or health risk to humans or animals, if ingested. These values are suitable for sediments in contact with water and serve as appropriate standards for the alternative use, no-closure option. Usually treatment or management of the *most limiting parameter* brings all other limiting parameters within the acceptable criteria. Debris, asphalt or other obvious sidewall contaminants are removed for aesthetic reasons under the no-closure minimum management option. The pit may require some stabilization of the berms by establishing a vegetative cover. Adequate stabilization results from a light application of fertilizer and hay mulch cover, followed by direct seeding or sprigging of native grasses into the mulch.

TABLE 9.1
GLC VALUES FOR MINIMUM MANAGEMENT

Parameter	Criteria
pH, s.u.	6–8
EC, mmhos/cm	4
ESP, %	15[a]
Total Metals, mg/kg	
Arsenic, As	40
Barium, Ba	40,000
Cadmium, Cd	10
Chromium, Cr	500
Copper, Cu	750
Lead, Pb	300
Mercury, Hg	10
Nickel, Ni	250
Selenium, Se	10
Zinc, Zn	500
Organics	
TPH, %	[a]

[a]ESP can be calculated from saturated paste SAR value.

9.5.2   No-Closure Option Considerations

- **Advantages**

  - Simple process
  - Can be performed at any time
  - Cost effective

- **Limitations (No-Closure Option)**

  - Landowner approval
  - Pit decommissioned, but still open
  - Equipment and manpower requirements
  - Earth moving equipment
  - Tractor to spread and disk in fertilizer
  - Tractor to spread and crimp mulch application
  - Crew to apply amendments, seed or sprig grasses at small sites

- **Costs**

  - Testing
  - Amendments
  - Fill material
  - Equipment (earth moving, farm tractor and implements)
  - Revegetation
  - Permitting and closure documentation

9.6   BURIAL

9.6.1   Basic Technology

The burial method of pit solids disposal relies on pit materials remaining a minimum of 1.5 m (5 ft) below grade and back filled with clean berm and fill soil, as needed, to create a slightly mounded condition. Burial can be accomplished in the original excavation, in trenches adjacent to the pit, or in an off-site area excavated for the purpose. Groundwater is a concern and should be at least 1.5 m (5 ft) below the excavation floor if, waste solids analyses show EC >16 mmhos/cm and the ESP <15%. Research shows pit solids do not give up salt at an EC <24 mmhos/cm and ESP >15%.

Waste materials are air-dried or chemically dried prior to burial to prevent dewatering into the surrounding soil matrix caused by backfill overburden pressure. Soluble constituents, in materials buried too wet, tend to rise by capillary action to the surface. Pit solids containing relatively fresh petroleum hydrocarbon must be dried with oil sorbent to prevent leakage from overburden pressure. For trenched wastes, oil sorbents generally are blanket applied above and below the buried pit solids. Lesser amounts of sorbents are effective when blended with the pit solids before placement, but this requires equipment and operator time to perform the blending. Low pH pit solids should be adjusted up to pH 6 with agricultural limestone prior to burial to maintain metals in a fixed state. Dust the top of the buried solids with gypsum at a rate of 536 kg/ha (500 lb/acre) prior to backfill operations.

Waste burial is best suited for small to moderately sized (<1,000 m², or 10,760 ft²) pits. A vegetative cover is required to prevent erosion exposing buried pit solids.

TABLE 9.2
GLC VALUES FOR BURIAL

Parameter	Criteria
pH, s.u.	6–9
EC, mmhos/cm	16–24
ESP, %	NA[a]
Total metals, mg/kg	
Arsenic, As	40
Barium, Ba	100,000
Cadmium, Cd	10
Chromium, Cr	1500
Copper, Cu	750
Lead, Pb	300
Mercury, Hg	10
Nickel, Ni	750
Selenium, Se	10
Zinc, Zn	1500
Organics	
TPH, %	3

[a]NA refers to not applicable.

9.6.2   Applicability of GLC Values

Burial GLC values (Table 9.2) assume a farming or potential future residential land use. The distribution of salt as sodium is not a concern for buried pit solids because materials are buried below the plant root zone. The barium criterion is elevated to 100,000 mg/kg when the source can be confirmed as barite (toxicity characteristic leaching procedure[1] (TCLP) for barium is <100 mg/L). Total chromium, lead, nickel and zinc are increased because materials are placed out of surface contact and below the active root zone. TPH criterion of 5% will require oil sorbent if the hydrocarbon can be released as a liquid under the nominal pressure (335 kPa, or 50 psi) required in TCLP test protocol.

9.6.3   Burial Option Considerations

- Advantages

  - Simple process
  - Can be performed at any time
  - Equipment readily available

- Limitations

  - Groundwater should be at least 1.5 m (5 ft) below land surface
  - Burying stops biodegradation of petroleum residuum
  - The only treatment is drying liquid fractions
  - Requires backfill with clean soil and vegetative cover

---

[1]SW-648 Method 1311

- **Equipment and Manpower Requirements**

  - Earth moving equipment
  - Experienced backhoe or trackhoe operator
  - Farm equipment to apply and distribute amendments
  - Tractor to spread and crimp mulch application
  - Crew to apply amendments, seed or sprig grasses at small sites

- **Costs**

  - Testing
  - Amendments
  - Fill material
  - Transportation, if buried at another location
  - Equipment (earth moving, farm tractor and implements)
  - Revegetation
  - Permitting and closure documentation

9.7 DILUTION BURIAL

9.7.1 Basic Technology

Dilution burial relies on mixing adjacent soil with pit waste solids to reduce constituent levels below GLC values (Table 9.3) and the resultant waste/soil mix is then buried a minimum depth of 0.9 m (3 ft) below grade and covered with non-admixed soil from the adjacent area or other fill material. Dilution burial is performed in the original pit excavation, in adjacent trenches or in an off-site location in an excavation constructed for this purpose. The material mix can occur at the surface and can be in contact with soil water, if the resulting blend meet or exceeds land treatment GLC values (Table 9.3). Pit solids must be adjusted to pH $\geq$ 6 with agricultural limestone prior to dilution burial to maintain metals in a fixed state.

Although the technique implies simple dilution, there are chemical alterations occurring to treat waste solids constituents. The primary advantage obtained from using dilution burial is the need for less land area for treatment/disposal. GLC values (Table 9.3) are less stringent than minimum management values, because materials are reacted with diluent soil and buried out of the active root zone. Revegetation is required to stabilize the land surface.

Prior to covering the diluted waste, the top of the admixed waste solids is dusted with agricultural gypsum at a rate of 560 kg/ha (500 lb/acre).

9.7.2 Applicability of GLC Values

Dilution burial is favored in agronomic land resource areas. The treatment occurs below the potential root zone, therefore salt and TPH values are managed at higher levels than the minimum management GLC guideline (Table 9.3). Sodium distribution (ESP) is not a factor. The organic biodegradation mechanism is restricted because of the depth of burial, but proceeds slowly. The recommended SPEC threshold for the resultant waste/soil mixture is 12 mmhos/cm., if the admixed waste is covered with 3 ft of clean soil. The recommended TPH value is 3%. Total barium, chromium, lead, nickel and zinc are relaxed relative to minimum management criteria, because of placement of the amended waste below the active root zone.

TABLE 9.3
GLC VALUES FOR
DILUTION BURIAL

Parameter	Criteria
pH, s.u.	6–9
EC, mmhos/cm	12
ESP, %	NA[a]
Total metals, mg/kg	
Arsenic, As	40
Barium, Ba	100,000
Cadmium, Cd	10
Chromium, Cr	1500
Copper, Cu	750
Lead, Pb	300
Mercury, Hg	10
Nickel, Ni	750
Selenium, Se	10
Zinc, Zn	1500
Organics	
TPH, %	3

[a]NA refers to not applicable.

9.7.3   Dilution Burial Option Considerations

- **Advantages**
  - Simple process
  - Can be performed at any time
  - Does treat waste constituents
  - Cost effective

- **Limitations**
  - Groundwater is considered at higher GLC values
  - Biodegradation of petroleum residuum restricted
  - Requires backfill with clean soil and vegetative cover

- **Equipment and Manpower Requirements**
  - Earth moving equipment
  - Tractor to spread and disk-in fertilizer
  - Tractor to spread and crimp mulch application
  - Crew to apply amendments, seed or sprig grasses at small sites

- **Costs**
  - Testing
  - Amendments
  - Fill material
  - Equipment (earth moving, farm tractor and implements)
  - Revegetation
  - Permitting and closure documentation

9.8   LAND TREATMENT

9.8.1   Basic Technology

Land treatment, also known as land farming or land spreading, is a method of *treatment* and disposal rendering pit solids innocuous by reaction with amendments and surface soil. This method uses dilution, chemical alteration, and biodegradation mechanisms to reduce constituents to acceptable soil levels commensurate with intended *non-waste* land use. The technology has been utilized for many years and recognized by the EPA (1995) as a preferred method for disposal of municipal sewage sludge and many nonhazardous industrial wastes.

The process entails mixing pit solids/impacted soils with sufficient clean soil so the final mix is at or below the parameter GLC values (Table 9.4). Mix ratios are usually calculated based on a 15-cm (6-in.) of receiving (background) soil layer of known composition. Loading rates are calculated based on the most limiting constituent. Fertilizer amendments are added to encourage biodegradation of petroleum hydrocarbon and to establish a vegetative cover. Blanket gypsum applications are incorporated into the waste/soil treatment zone to correct excess sodium normally associated with pit waste solids.

The pit is closed after the waste solids are removed and spread on the receiving soil. Berm material is used as the initial backfill followed by treated waste/soil mix or topsoil, as needed, to restore the pit to the original grade and contour. *The pit bottom can be utilized as the receiving soil, if property land area is small.*

9.8.2   Applicability of GLC Values

Land treatment GLC values used for pit closures assume agricultural (pasture/forage grass) land use. These criteria parallel minimum management GLC values, with the exception of SPEC. SPEC is higher (8 mmhos/cm compared to 4 mmhos/cm) for the land treatment process, because waste solids are restricted to surface application into a thin (15 cm) receiving soil layer. There is no limitation to

TABLE 9.4
GLC VALUES FOR LAND TREATMENT

Parameter	Criteria
pH, s.u.	6–8
EC, mmhos/cm	4–8
ESP, %	<15%
Total Metals, mg/kg	
Arsenic, As	40
Barium, Ba	100,000
Cadmium, Cd	10
Chromium, Cr	1500
Copper, Cu	750
Lead, Pb	300
Mercury, Hg	10
Nickel, Ni	750
Selenium, Se	10
Zinc, Zn	1500
Organics	
TPH, %	3

depth of materials at SPEC 4 mmhos/cm. Pit solids high in TPH concentration are often applied at loading rates considerably >1% GLC and biodegraded with time to clean closure standard. Biodegradation is best accomplished during dry season with irrigation water applied as needed. Typically, irrigation water is applied at 1 in. of water per week.

9.8.3 Land Treatment Option Considerations

- **Advantages**

  - Simple efficient process
  - Does treat waste constituents
  - Can be used to dispose of large volumes at moderate contamination
  - Effective in treatment and disposal of relatively high TPH wastes
  - Cost effective

- **Limitations**

  - Requires relatively large treatment area
  - Biodegradation requires amendments, aeration and water.
  - Mechanical and labor inputs needed on follow up basis

- **Equipment and Manpower Requirements**

  - Earth moving equipment (grader, backhoe, front-end loader)
  - Tractor to spread and disk in fertilizer
  - Tractor to spread and crimp mulch application
  - Crew to apply amendments, seed or sprig grasses at small sites
  - Fertilizer nutrients (nitrogen, phosphorus, potassium) and gypsum

- **Costs**

  - Testing
  - Amendments
  - Fill material
  - Equipment (earth moving, farm tractor and implements)
  - Revegetation
  - Permitting and closure documentation

9.9 IN SITU LEACH FOR SALINE PIT SOLIDS

9.9.1 Basic Technology

In situ leach is the process of reducing high-salt levels by normal leach mechanisms using rain precipitation or suitable quality irrigation water. This process is conducted in a pit environment for control and recovery of salt-water leachate. Pit solids are first treated with gypsum to displace sodium on the exchange complex. Also, gypsum is necessary to prevent solids dispersion on leaching excess salt. Natural organic materials not readily degraded (corn stalks, wood chips, straw, etc.) are mixed into the pit solids to enhance porosity and movement of water through the waste solids. Waste solids are configured within the pit into windrows with furrows draining by gravity to a common sump at the low end of the pit floor. Windrows provide a leaching profile for the altered (treated with gypsum and mixed with organic fill) waste solids.

The sump is constructed below the pit floor to facilitate rapid drainage from the furrows and is lined with plastic to control seepage. The leachate is removed from the sump on as-needed basis for disposal into the salt-water injection system.

In situ leach is considered a passive process, because it depends on a natural mechanism, and uncontrolled distribution of rainfall over several years. Rainfall must exceed the evaporation rate of the bare soil to be effective. For this reason, we can expect the leach process to function during the rainy season with 20–30% of the salt leached from the waste each season. Amendment applications and maintenance such as turning windrows is done during the dry season.

In situ leach is a treatment process used to lower salt concentrations in waste solids for on-site disposal. The treatment is expected to take from 5 to 7 years, with final disposal using dilution burial (Table 9.3) or land treatment techniques (Table 9.4). Waste solids are monitored on a yearly basis and the analytical results used to adjust amendments. Also, the EC, ESP and TPH results are used to make decisions relative to disposal and final pit closure using the appropriate GLC values.

After remediation is complete, berm material and low-salt fill is used to establish the normal grade and contour. A tease application of fertilizer and hay mulch cover, with direct seeding or sprigging of native grasses into the mulch will complete the pit closure.

9.9.2   Applicability of GLC Values

Appropriate GLC values depend on the final disposal and pit closure method. Ordinarily dilution burial (Table 9.3) is the disposal method of choice; when native soils underlying pit solids are relatively low in salt, SPEC <8 mmhos/cm. If underlying soil is high in salt (EC >16 mmhos/cm) we recommend the leach process continue until the waste solids meet the land treatment EC criteria of 8 mmhos/cm. If underlying soil is relatively low in salt (<8 mmhos/cm) then the waste solids are treated by the leach process to EC < 16 mmhos/cm and mixed with underlying soil to effect disposal by dilution burial. The target EC for treatment can be as high as EC 32 mmhos/cm if treated pit solids can be effectively diluted and readily mixed with underlying low-salt native soil.

9.9.3   In Situ Leach Option Considerations

- **Advantages**

  - Treats large volume of high-salt wastes
  - Utilizes pit construction for confinement
  - Passive process is cost effective

- **Limitations**

  - Multiple year process
  - Requires mechanical manipulation of large volume of materials
  - Requires large volume of organic matter

- **Equipment and Manpower Requirements**

  - Earth moving equipment (caterpillar and track-hoe)
  - Tractor to spread and disk in fertilizer
  - Tractor to spread and mix organic matter
  - Organic amendments
  - Gypsum
  - Low-salt backfill
  - Fertilizer amendments

- Costs

  - Testing
  - Amendments
  - Fill material
  - Equipment (earth moving, farm tractor and implements)
  - Revegetation
  - Permitting and closure documentation

9.10 IN SITU BIOREMEDIATION FOR OILY WASTE SOLIDS

9.10.1 Basic Technology

In situ bioremediation is a controlled treatment process utilizing naturally occurring soil bacteria to degrade petroleum hydrocarbons into humus, carbon dioxide and water. The application of this method is controlled in a pit environment with the incorporation of large amounts of organic bulking agents such as wood chips, corn stalks, straw, dried animal manure, saw dust, leaves, paper or other organics. The bulking agents serve as absorbent, co-metabolites, and nutrients. In addition, the bulking agents increase water holding capacity, porosity and aeration of the waste. Also, the bulking agents dilute the TPH and salt concentrations of the waste.

In practice the organic matter and fertilizer amendments are mixed together, followed by construction of windrows within the pit to facilitate aeration and the application of irrigation water on an as-needed basis. Windrows are turned periodically by mechanical means in order to increase exposure to oxygen and maintain necessary porosity.

Bioremediation of pit solids is continued until the TPH level < 10%, at which point the pit solids are disposed of by a more conventional land treatment, dilution burial or burial technique. Uncontaminated berm material or other clean soil are used to backfill the pit to effect closure, if a dilution burial or burial technique is employed. If materials are degraded to the land treatment GLC value of 1% there would be no restriction to use treated solids as fill.

9.10.2 Applicability of GLC Values

Appropriate GLC values are dictated by the pit solids disposal and final pit closure method selected. We recommend in situ bioremediation continue until pit solids meet the burial TPH criteria (Table 9.3). At this point the pit solids can be disposed of by burial or dilution burial technique. Berm or levee soils are used as the initial backfill, followed by uncontaminated fill or admix of waste solids and clean soil to land treatment GLC criteria.

9.10.3 In Situ Bioremediation Considerations

- **Advantages**

  - Less land area required than other treatment processes
  - Potentially shorter treatment duration
  - Materials handled on site
  - Cost effective

- **Limitations**

  - Requires significant amendment input

- Weekly irrigation during dry season
- Turning windrows at 3- to 4-week intervals

- **Equipment and Manpower Requirements**

  - Earth moving equipment
  - Tractor to spread and disk in fertilizer
  - Tractor to spread and crimp mulch application
  - Crew to apply amendments, seed or sprig grasses at small sites

- Costs

  - Testing
  - Amendments
  - Fill material
  - Equipment (earth moving, farm tractor and implements)
  - Revegetation
  - Permitting and closure documentation

9.11 SOLIDIFICATION/CHEMICAL STABILIZATION

9.11.1 Basic Technology

Solidification and chemical stabilization are treatment technologies encapsulating and immobilizing waste materials for safe disposal or reuse. These techniques are used for one or a combination of the following purposes:

- Improve handling and/or physical characteristics of the waste
- Limit solubility of waste constituents
- Decrease surface area from which the loss of constituents can occur
- Detoxify a waste by neutralization with lime

Solidification is the process of adding a sufficient quantity of solidifying agent (e.g., cement) to produce a solid mass of material, which has high structural strength and low permeability. Contaminants may or may not react chemically with the waste, but in all cases the waste is locked within the matrix. The solidified mass can be used as construction fill, simply buried on site or removed for burial off site.

Chemical stabilization is a process designed to structure waste solids into a soil-like constituency. Usually this is accomplished with a drying process, which also reduces constituent solubility or detoxifies the admixed waste. This can entail neutralization of strongly acid reactions with lime.

A third application for this process involves the beneficial reuse of E&P waste solids in the construction of drill sites (e.g., drilling pads) and the construction and/or repair of lease roads. The difference between this process and road spreading is that waste solids are reacted with cementitous agents or pozzolans chemically and physically bind the matrix.

9.11.2 Applicability of GLC Values

The application of specific criteria for solidification treatment is largely dependent on the final disposal method to be used. In some instances, particularly for oily pit solids it may be necessary to manage fluidity prior to TCLP testing. Liquids extracted under pressure are tested neat under TCLP protocol. This greatly increases the potential of oily wastes failing the test. The following options are available for solidification:

TABLE 9.5
LEACHATE TEST CRITERIA

Parameter	Criteria
pH, s.u.	6–12
Chloride, mg/L	< 500
SAR, unitless	NA[a]
Total metals, mg/kg	
Arsenic, As	< 0.5
Barium, Ba	< 10.0
Cadmium, Cd	< 0.1
Chromium, Cr	< 0.5
Lead, Pb	< 0.5
Mercury, Hg	< 0.02
Selenium, Se	< 0.1
Zinc, Zn	< 5.0
Organics	
TPH, %	< 10.0

[a]NA refers to not applicable.

- Excavation and partial immobilization of the waste is used for improving material handling characteristics or reducing fluidity prior to disposal. Partial immobilization is often done to reduce solubility of constituents in water, rendering materials acceptable for on site disposal.
- Excavation and nearly complete immobilization of the waste is done to convert wastes to a solid mass with little potential for the release of soluble constituents or fluid components. Treated materials can be recycled as construction fill or buried in the pit excavation. If solidified material is to be buried, the bottom of the burial cell must be at least 1.5 m (5 ft) above the water table and the top of the mass should be at least 1.5 m (5 ft) below the land surface. Backfill material must meet GLC criteria for minimum management.
- Solidification and chemical stabilization are treatments reserved for RCRA-exempt E&P wastes having high petroleum hydrocarbon and/or salt concentrations, and can be modified to meet strict reuse criteria. Solidified materials are subjected to a 1:4 (wastewater) leachate test to evaluate soluble constituent concentrations criteria given in Table 9.5. Metals are analyzed for TCLP extract using EPA Method 1311 or its equivalent.

Solidified material should also meet the following physical criteria:

- Unconfined compressive strength   >200 kPa
- Permeability $<1 \times 10^{-6}$ cm/sec
- Wet/dry durability   >10 cycles to failure

9.11.3 Solidification Option Considerations

- **Advantages**
  - Raw materials for the process are readily available
  - Conventional equipment available
  - Used to convert hazardous wastes to nonhazardous
  - Option for treating problematic mixed wastes (salts and organics)

- **Limitations**

    - Ultimate destruction of the compounds generally does not occur
    - Large quantities of cement and other agents often required
    - Costs are significant, but could be only option for on-site disposal

- **Equipment and Manpower Requirements**

    - Chemical storage facilities
    - Materials handling equipment
    - Materials mixing equipment
    - Treated materials handling equipment
    - Two-to-three operator participants are needed for small jobs
    - As many as 10 operator participants may be required for larger sites

- **Costs**

    - Testing
    - Solidification/stabilization additives
    - Fill material
    - Labor
    - Disposal of treated solids
    - Permitting and closure documentation

9.12   ROAD SPREADING

9.12.1   Basic Technology

Road spreading manages waste for beneficial use wherein weathered oil contaminated drilling fluids, sludges/soils or weathered matted oil residues are mixed with an existing road surface for repair or blended with road mix materials. The resulting combined mix is spread over the lease roads as a riding surface. The most suitable application is for low-salt, high organic waste solids with an acceptable flashpoint (<140°F). Road spreading is ideally suited for the beneficial use of weathered asphaltic base oil not amenable to biodegradation. This waste should be consider a resource and recycled for road repair and construction, whenever possible.

Wastes from fresh oil spills and leaks are best managed by blending with weathered oil contaminated waste/soil prior to road application. Weathered materials have a capacity to absorb lower viscosity hydrocarbon with greater fluidity. Also, the road spreading process, controls light-end components, which can partition into an aqueous phase during storm water runoff. We find the light-end fraction of crude oil in soil approaching the residual saturation is stabilized with the addition of 20% by weight of hydrated agricultural limestone.

9.12.2   Road Spreading Procedure

First prepare the road surfacing materials by removing all free oil or water. Materials are blended together within the containment facility and tested for the GLC parameters (Table 9.6). If the materials are within the criteria, they can be used without modification as road spreading material. Gravel is added in approximate equal volume to provide the porosity for the blend to fill in forming a riding surface. Gravel is not always necessary for resurfacing an existing road, or in preparing the initial base coat.

Next prepare the roadway so the surfacing material can be applied to the receiving roadbed. In preparation for this application, the roadbed is scraped, turning loosened material to each side of the road providing parallel berms or trough for the application of road surfacing materials. Gravel is

TABLE 9.6
GLC VALUES FOR ROAD
SPREADING

Parameter	Criteria
pH, s.u.	6–10
EC, mmhos/cm	< 12
ESP, %	NA[a]
Total Metals, mg/kg	
Arsenic, As	40
Barium, Ba	100,000
Cadmium, Cd	10
Chromium, Cr	1500
Copper, Cu	750
Lead, Pb	300
Mercury, Hg	10
Nickel, Ni	750
Selenium, Se	10
Zinc, Zn	1500
Organics	
TPH, %	NA
Flashpoint, °C/°F	< 60/140

[a]NA refers to not applicable.

added to the trough if the initial scraping reached subsurface clay. This is followed by the application of oily waste solids.

Salts are effectively managed by blending with existing road surface materials at controlled mix ratios or mixing with low-salt waste materials prior to road application. Waste oil solids are graded and allowed to dry if wet. The edge berms are moved to the center forming one row. This row is rolled flat over the entire road followed immediately by a packing operation to complete the application. Packing is perhaps the most critical step and should extend to the road edges. Crown the road to prevent surface pooling of water during rainfall events.

Take precautions to reduce potentially adverse impacts to environmentally sensitive areas, such as watercourses, by doubling the berm heights, and removing any pooled water from the road using a vacuum truck. Visual monitoring for a no-sheen runoff from the roadway is an additional quality control measure, costing very little, but serving greatly to protect the environment.

9.12.3 Applicability of GLC Values

The main concerns regarding road spreading are associated with surface-wear dust and runoff. For this reason, we recommend the operator adhere to the management constraints set forth in the GLC Values for Road Spreading (Table 9.6).

9.12.4 Road Spreading Option Considerations

- **Advantages**
  - Provides usable material
  - Disposal option for high TPH wastes not suited for biodegradation
  - Cost effective

- **Limitations**

  - Possible problem with flash point, salinity and metal content
  - Oily wastes need fillers and aggregate to be useful

- **Equipment and Manpower Requirements**

  - Earth moving equipment (graders, front end loaders, backhoes)
  - Mixing equipment
  - Manpower to operate equipment

- **Costs**

  - Testing
  - Labor
  - Heavy equipment
  - Mix materials
  - Permitting and closure documentation

9.13   THIRD-PARTY DISPOSAL

9.13.1   Basic Technology

Third-Party Disposal involves an outside (company-approved) third-party contractor accepts company-generated wastes at a properly permitted facility where approved storage or remediation methods are used. Keep in mind; in the United States the generator of waste remains responsible for the waste. The third party disposer is only a custodian of the company waste, unless the waste is *converted to a nonwaste material.*

9.13.2   Applicability of GLC Values

Third-Party Disposal generally is used when most other options are not available. Ordinarily this is the most expensive option, because of costs associated with waste removal, waste transport to an off-site location, and disposal charge by the third-party. Wastes are removed from the pit so there are no applicable GLC criteria; accept as might be applied to off-site treated solids used as backfill.

- **Advantages**

  - Allows for immediate backfill of pit
  - Provides for clean closure and immediate land use

- **Limitations**

  - Third-party must be approved by the company prior to use
  - Loss of control over the waste disposal process
  - Potential future liability associated with the third-party operations. The third-party waste handler is only a custodian of waste. The generator retains custody of the waste.
  - Costs for waste disposal increased

- **Equipment and Manpower Requirements**

  - Earth moving equipment to excavate pit solids
  - Manpower to excavate site and backfill

- Tractor and implements used to revegetate site
- Crew to apply amendments, seed or sprig grasses at small sites

- Costs

  - Testing
  - Amendments
  - Fill material
  - Equipment (earth moving, farm tractor and implements)
  - Revegetation
  - Charges by third party to take and dispose of waste
  - Bill of lading documentation
  - Permitting and closure documentation

9.14   PIT CLOSURE DOCUMENTATION

9.14.1   Documentation Procedure

It is essential activities associated with the closing of drilling reserve and/or emergency pits be documented and filed to allow reconstruction of events for future company audits or suites. A file should be established for each pit consisting of a series of attachments documenting the following:

- Pit name
- Field/plant
- Pit location
- Pit permit number, if applicable
- Decommissioning and closure approval (permit)
- Laboratory analysis report for liquid and solids pit contents
- Field and site drawings
- Schematics showing sampling points, dimensions, liquid/solid volumes
- Photographs
- Landowner correspondence
- Any exemptions/exceptions/correspondence from Government Agencies.
- Data summarizing pit closure project:

  - Closure date:
  - Pit identification
  - Brief description of closure (cleaning, disposal, backfilling, etc.)
  - Contractor, disposal permits, volumes transported and to where
  - Analytical tests demonstrating closure compliance status
  - Costs

Before a "closed" status designation is assigned to a pit, documentation should be submitted to theAgency or Company Department authorizing the closure or decommissioning. The authorization form (Table 9.7), provided below presents a checklist of necessary information and the recommended review/authorization procedure.

Once the pit has been closed or decommissioned for alternate beneficial use, the field personnel must send the completed checklist and all supporting documentation to Engineering or Environmental Issues/Regulatory Affairs personnel assigned the responsibility for maintaining and auditing the pit closure file.

9.14.2   Checklist and Authorization Form

The "PIT CLOSURE CHECKLIST AND AUTHORIZATION FORM," Table 9.7, serves as the cover sheet for the closure documentation package, and each attachment verifying specific closure activities must be identified with the pit permit number or appropriate company reference number.

TABLE 9.7
PIT CLOSURE CHECKLIST AND AUTHORIZATION FORM

Pit Name:	Location:	Field:		State:
Pit Permit No:	Reference #:	Review/Approval (Initials)		
		Field Supt.	Reg. Group	
1. Company/landowner correspondence				
2. Laboratory analysis of pit contents				
3. Instructions, pit sample points				
4. Site drawings				
5. Legal descriptions, plat, GPS coordinates				
6. Photographs (pre- and postclosure)				
7. Regulatory correspondence				
8. Summary information for pit closure				
a. Closure date:				
b. Pit ID: Name, location, field, reference #				
c. Closure details (brief description of method)				
d. Waste disposal records:				
- Volume disposed on site				
- Volume disposed off site				
- Disposal permits				
- Contractor manifests and trip tickets				
e. Copy of AFE				
9. Copy of pit permit if available				
10. Closure/alternate use certification				

_____   _____
Final Approval Authority (Name and Title)   Date Approved

REFERENCES

EPA, 1983. Hazardous Waste Land Treatment. Washington, D.C.: Office of Water Regulations and Standards, Criteria and Standards Division. SW-874.

EPA, 1995. "Pollution Limits," 40 CFR §503.13.

Freeman, B.D. and L.E. Deuel. 1984. "Guidelines for Closing Drilling Waste Fluid Pits in Wetland and Upland Areas". Presented at the 7th Annual Energy Sources Technology Conference and Exhibition. Sponsored by The Petroleum Division—ASME, New Orleans, Louisiana, February 12–16, 1984.

Rule 29-B, 2000. "Title 43, Part 19, Chapter 3, §§ 311, 313, and 315—Pollution Control, State of Louisiana, Department of Natural Resources, Office of Conservation", Baton Rouge, LA, Promulgated January 20, 1986; revised October–December 2000.

PROBLEM SET

10.1 INTRODUCTION

This chapter provides an opportunity to solve practice remediation problems taken from actual field situations. The problem solutions are provided near the end of the chapter. Problem sets #4, #5 and #6 use the data listed in Table 10.1.

10.2 PROBLEMS

10.2.1 Problem #1

The site is in an oilfield located near Kingsville, TX. An oil surface run-down line sprang a leak from a corroded section of the line. Based on the production rate, the foreman estimated 105 barrels of 35° API crude oil leaked onto the soil. The spill occupied a 0.25-acre area, with oil penetration to a depth of 4 in.

1. What is the initial soil loading?
2. What is the allowable loading in Texas?
3. How much background soil is required to bring the spilled oil to the 5% maximum TPH loading rate? Assume the background soil contains 0.01 percent TPH.
4. How much nitrogen is required to bioremediate the carbon added by the crude oil spill?
5. How much fertilizer is required? How much P&K would you add, assuming you had ammonium sulfate (21% N), super phosphate (21% P_2O_5; 9.2% P), and muriate of potash (61% K_2O, 51% K) available?
6. How would you apply the fertilizer?

10.2.2 Problem #2

The Johnson #1 site is located in Wyoming where the annual mean temperature of the soil is 15°C. A flowline leaked 15 barrels of 20° API oil and 110 bbl of produced water onto the soil. The spill area covered an area of 0.1 acres 1 in. deep, and the soil type in the spill area is clay.

1. What is the initial oil loading?
2. Is this loading acceptable for the temperature conditions?
3. Measurements in the field using a 1:1 by weight showed an EC 12 mmhos/cm. What is the saturated paste EC (SPEC) for the contaminated soil?
4. Is the SPEC satisfactory, or must the operator remediate the spill area?
5. The background SPEC is 1.0 mmhos/cm. Could the area be remediated by simple dilution? If so, how much land would be required to meet Wyoming regulations?

TABLE 10.1
ANALYTICAL DATA FOR SMITH WELL NO. 4

Parameter	Reserve Pit	Shale Pit	Oil Pit	Bkgd Soil	Tech	Date
Moisture, %	35.1	27.2	42.0	12.0	KM	12/04
SP Moisture, %	98.4	47.7	61.3	27.2	KM	12/04
pH, s.u.	9.4	10.8	11.5	7.8	JEF	12/04
SPEC, mmhos/cm	62.2	13.2	10.2	0.8	JEF	12/04
1:1 Chlorides, ppm	18,671	1,843	1,392	46.0	JEF	12/05
1:1 Soluble Cations, mEq/L						
Sodium	514.3	44.0	29.2	2.5	ALB	12/07
Calcium	38.3	8.7	10.1	3.1	ALB	12/07
Magnesium	4.4	1.3	0.1	1.7	ALB	12/07
Potassium	10.5	9.4	10.6	0.7	ALB	12/07
SAR, unitless	111.3	19.7	13.0	1.6	Calculation	
CEC, mEq/100 g	22.6	8.0	5.9	6.1	ALB	12/12
Exchangeable cations, mEq/100 g						
Sodium	17.8	0.6	1.0	0.4	ALB	12/12
Calcium	75.2	67.1	155.2	27.1	ALB	12/12
Magnesium	0.7	1.3	1.8	1.1	ALB	12/12
Potassium	1.7	2.4	1.9	1.2	ALB	12/12
ESP, %	78.9	7.2	16.9	6.6	Calculation	
Total metals, mg/kg						
Arsenic, As	5.0	4.5	5.0	0.6	ALB	12/14
Barium, Ba	8,210	1,820	568	225	ALB	12/15
True total Ba	13,867	4,117	1,190	566	ALB	12/15
Cadmium, Cd	1.4	1.1	1.6	0.9	ALB	12/15
Chromium, Cr	13.0	18.0	32.0	2.1	ALB	12/15
Lead, Pb	18.0	24.0	49.0	9.6	ALB	12/15
Mercury	0.2	0.4	< 0.1	0.1	ALB	12/14
Selenium, Se	0.5	0.4	0.4	0.1	ALB	12/14
Silver, Ag	2.0	5.0	5.0	1.6	ALB	12/15
Zinc, Zn	67.3	61.8	67.1	42.3	ALB	12/15
Oil & Grease, %	1.3	8.5	7.2	< 0.1	JEF	12/08

6. Could the land be treated with fertilizer and gypsum? How much fertilizer and gypsum would be required? Assume the soil has a CEC of 50 meq/100g and an ESP of 59 percent.
7. Describe the fertilizer and gypsum application procedure.

10.2.3   Problem #3

A water flood storage tank is located in a wheat field. A pipeline leak causes 210 bbl of produced water to impact a vegetated area. Sampling and analysis show the spill area SPEC 62.2 mmhos/cm, CEC 25 meq/100g, ESP 67% and the pH 9.3. The leak affected 0.2 acres land to a depth of 6 in. Using the above data, calculate the following values:

1. What is the resulting TDS of the soil?
2. Can this site be remediated with only gypsum and sulfur?

3. Assuming the spill contaminated 0.2 acres of farmland to a depth of 6 in., how much gypsum and sulfur would be required to return the soil for wheat production?
4. How much land is required to achieve a treatment area having SPEC 6.0 mmhos/cm, assuming the native soil has a SPEC of 1.0 mmhos/cm?
5. How deep should one spread the contaminated soil?
6. What steps should one take to land treat site?

10.2.4 Problem #4

Smith Well No. 4 is located on a farm near Kingsville, TX. The *reserve* pit (Table 10.1) is 100 ft by 150 ft and the drilling fluids are approximately ft deep.

1. Calculate the 1:1 EC (mmhos/cm).
2. Compare the 1:1 EC to the soluble cations and anions and show your calculations. Explain the results. (Hint: meq/l Cl = mg/L Cl ÷ 35.5 mg/meq).
3. Which parameters exceed the regulatory thresholds under Texas regulations for landspreading?
4. Are there any other parameters, which should be considered to insure the land is returned to full agronomic potential? The land should be returned to a level at which sorghum will grow. (Hint: See Plate 12.)
5. a) Calculate the volume of soil needed for dilution of the limiting constituent for landspreading. You may have to calculate several parameters to determine which is the most limiting.
    b) Determine the amount of land necessary for landspreading.
    c) Calculate the application depth of the drilling muds for landspreading
    d) Calculate the application rates of any necessary amendments.
    e) Write up recommendation for the landspreading application.
    f) Calculate land requirement to spread the Smith Well #4 reserve pit solids in Oklahoma.

10.2.5 Problem #5

Smith Well No. 4 is located in an elevated freshwater wetland near Baton Rouge, LA. The *shale* pit (Table 10.1) is 75 ft x 50 ft and the drilling muds and cuttings are approximately 5 ft deep.

1. What are the appropriate 29-B criteria for pit closure in elevated wetlands in Louisiana?
2. Calculate the 1:1 EC (mmhos/cm):
3. Compare the 1:1 EC to the soluble cations and anions and show your calculations. Explain the results. (Hint: 1meq/l Cl = 35.5 ppm Cl)
4. Which parameters exceed the regulatory thresholds under Louisiana regulations for land treatment?
5. Are there any other parameters, which should be considered to ensure the land is returned to its full potential? The land should be returned to its full agronomic potential.
    a) Calculate the volume of soil needed for dilution of the limiting constituent for landspreading. You may have to calculate several different items to determine which is the most limiting.
    b) Determine the amount of land necessary for landspreading.
    c) Calculate the application depth of the drilling muds for land treatment.
    d) Calculate the application rates of any necessary amendments.
    e) Write up recommendations for the landspreading application.

10.2.6 Problem #6

Smith Well No. 4 is located at a farm near Tulsa, OK. The *shale* pit (Table 10.1) is 75 ft x 50 ft and the drilling muds are approximately 5 ft deep.

1. Calculate the wet weight of pit solids in lb/gal.
2. Calculate the dry weight percentage of the solids.
3. Calculate the volume of solids in the pit in barrels.
4. Using OCC calculations (OAC 165:10 Appendix 1 of rule), determine the acreage required for landspreading of solids using the dry weight application limitation (200,000 lb/acre dry solids).
5. Calculate the ppm TDS in the receiving soil (Bkgd).
6. Calculate the mg/L TDS of the pit solids.
7. Using OCC calculations (OAC 165:10 Appendix 1 of rule), determine the acreage required for landspreading of solids using the total dissolved solids application limitation (6,000 lb. TDS/ acre).
8. Calculate the ppm Oil & Grease in the pit solids
9. Using OCC calculations (OAC 165: 10 Appendix 1 of rule), determine the acreage required for land spreading of solids using the oil & grease (O&G) application limitation (40,000 lb O&G/acre).
10. Based upon the OCC regulations for land application of solids, what is the minimum land requirement for disposal of these pit solids?

10.3   SOLUTION TO PROBLEMS

10.3.1   Solution to Problem #1

The site is in an oilfield located near Kingsville, TX. An oil run-down line sprang a leak from a corroded section of the line. Based on the production rate, the foreman estimated 105 barrels of 35° API crude oil leaked onto the soil. The spill occupied on 0.25-acre area, with oil penetration to a depth of 4 in.

1. What is the initial soil loading?

   Density of 35° API oil = 141.5/(131.5 + 35) = 0.850 g/ml = 0.85 kg/l
   Determine weight of oil spilled:
   105 bbl × 42 gal/bbl × 3.8 L/gal × 0.85 kg/l × 2.2 lb/kg = 31,337 lb oil
   Determine weight of contaminated soil:
   1 acre-6 in. = 2,000,000 lb soil; 1 acre 4 in. = (2,000,000/6) × 4 = 1,333,333 lb soil/acre-4 in.
   1,333,333 lb soil/acre-4 in. × 0.25 acre-4 in. = 333,333 lb soil
   Determine oil-loading rate
   (31,337 lb oil / 333,333 lb soil) × 100 = 9.4% oil

2. What is the allowable loading in Texas?

   Under Rule 91, 5% maximum TPH, with remediation to 1% within 1 year

3. How much background soil is required to bring the spilled oil to the 5% maximum TPH loading rate? Assume the background soil contains 0.01% TPH.

   Determine volume of contaminated soil:
   0.25 acres × 43,560 ft²/acre × 4 in. depth/12 in./ft = 3,630 ft³ contaminated soil

   Calculate native soil required to achieve 5% oil in treatment zone:

   $$5\% \text{ oil} = \frac{(3630 \text{ ft}^3 \text{ contaminated soil} \times 9.4\% \text{ oil}) + (X \text{ ft}^3 \text{ native soil} \times 0.01\% \text{ oil})}{(3630 \text{ ft}^3 \text{ contaminated soil} + X \text{ ft}^3 \text{ native soil})}$$

Solve for X:
$$(3630 \text{ ft}^3 + X \text{ ft}^3) \times 5 = (3630 \text{ ft}^3 \times 9.4\%) + (X \text{ ft}^3 \times 0.01\%)$$
$$4.99X = 15,972$$

X = 3200 ft^3 of native soil required to dilute to 5% oil in the treatment area.

Total treatment volume = contaminated volume + dilution volume
$$= 3630 \text{ ft}^3 + 3200 \text{ ft}^3$$
$$= 6830 \text{ ft}^3$$

Or:

Dilution volume = $\dfrac{[(\text{measured O\&G}) - (\text{target O\&G})] \times \text{contaminated volume, ft}^3}{[\text{target O\&G} - \text{Bkgd O\&G}]}$

$$= \frac{[(9.4\% - 5.0\%)] \times 3630 \text{ ft}^3}{[5.0\% - 0.01]\%}$$

X = 3200 ft^3

Treatment volume = $\dfrac{[(\text{measured O\&G}) - (\text{Bkgd O\&G})] \times \text{contaminated volume, ft}^3}{[\text{target O\&G} - \text{Bkgd O\&G}]}$

$$= \frac{[(9.4\% - 0.01\%)] \times 3,630 \text{ ft}^3}{[5.0\% - 0.01\%]}$$

$$= 6830 \text{ ft}^3$$

4. How much nitrogen is required to counteract the carbon added by the crude oil?

Determine the weight of carbon contained in the spilled oil
31,337 lb oil × 0.78 lb carbon/lb oil = 24,443 lb C

Determine the amount of nitrogen required, to achieve the desired C:N ratio of 150:1

$$\frac{24,443 \text{ lb C}}{N} = \frac{150}{1}$$

$$\frac{24,443 \text{ lb carbon} \times \text{lb N}}{150 \text{ lb C}} = 163 \text{ lb N required}$$

5. How much fertilizer is required? How much P&K would you add, assuming you had ammonium sulfate (21% N), superphosphate (21% P$_2$0$_5$; 9.2% P) and muriate of potash (61% K$_2$0, 51% K) available?

Determine nitrogen fertilizer requirement.

163 lb N required/0.21 lb N /lb ammonium sulfate = 776 lb ammonium sulfate

Determine the P & K required to achieve a 4:1:1 N:P:K ratio:

163 lb N required/4 lb N/lb of P or K = 41 lb P and 41 lb K required

Determine the phosphorus fertilizer requirement:

41 lb P required / 0.092 lb P/lb SUP = 446 lb superphosphate

Determine the potassium fertilizer requirement:
41 lb P required/0.51 lb K/lb muriate of potash = 80 lb muriate of potash

6. How would you apply the fertilizer?

Make split nitrogen applications of 1/2, 1/4 and 1/4 the total nitrogen fertilizer requirement
Apply all of the phosphorus and potassium with the initial nitrogen application.
Apply 388 lb ammonium sulfate, 446 lb superphosphate and 80 lb muriate of potash.
Wait 6–8 weeks, and then apply an additional 194 lb ammonium sulfate.
Wait 6–8 weeks, and then apply the final 194 lb ammonium sulfate.
Disk and cross-disk to a depth of 6 in. after each fertilizer application to incorporate amendments and aerate the treatment zone.

10.3.2   Solution to Problem #2

The Johnson #1 site is located in Wyoming, where the annual mean temperature of the soil is 15°C. A flowline leaks 15 barrels of 20° API oil and 110 bbl of produced water onto the soil. The spill area covers an area of 0.1 acres 1 in. deep, and the soil type in the spill area is clay.

1. What is the initial oil loading?

Density of 20° API oil = 141.5/(131.5 + 20) = 0.934 g/ml = 0.934 kg/L

Determine weight of oil spilled:
        15 bbl × 42 gal/bbl × 3.8 Ls/gal × 0.934 kg/L × 2.2 lb/kg = 4919 lb. oil

Determine weight of contaminated soil:
        1 acre-6 in. = 2,000,000 lb soil; 1 acre-1 in. = 2,000,000 /6 × 1 = 333,333 lb soil/acre-1 in.
        333,333 lb soil/acre-1 in. × 0.10 acre-1 in. = 33,333 lb soil

Determine the oil-loading rate:
        4919 lb oil/(33,333 lb. soil × 100) = 14.8% oil

2. Is this loading acceptable for the temperature conditions?

No, according to Table 8.2, for a mean annual soil temperature of 15° C, an acceptable loading rate would be 2% TPH.

3. Measurements in the field using a 1:1 by weight showed an EC of 12 mmhos/cm. What is the saturated paste EC (SPEC) for the contaminated soil?

Assume a moisture saturation percentage of 48% for a clay soil, and then calculate the SPEC from 1:1 EC:
        SPEC = (1:1 EC × 100)/% water at saturation
             = (12 mmhos/cm × 100)/48
             = 25 mmhos/cm

4. Is the SPEC satisfactory, or must the operator remediate the spill area?

The soil must be remediated to a SPEC of 4 mmhos/cm to meet Wyoming regulations.

5. The background SPEC is 1.0 mmhos/cm. Could the area be remediated by simple dilution? If so, how much land would be required to meet Wyoming regulations?

Yes, simple dilution is possible.

Determine volume of contaminated soil:
        0.10 acres × 43,560 ft^2/acre × 1 in. depth/12 in./ft = 363 ft^3 contaminated soil

Calculate the native soil required to achieve an SPEC of 4 mmhos/cm in treatment zone:

4 mmhos/cm = $\dfrac{(363\ ft^3\ contaminated\ soil \times 25\ mmhos/cm) + (X\ ft^3\ native\ soil \times 1\ mmhos/cm)}{(363\ ft^3\ contaminated\ soil + X\ ft^3\ native\ soil)}$

Solve for X:

$(363\ ft^3 + X\ ft^3) \times 4 = (363\ ft^3 \times 25\ mmhos/cm) + (X\ ft^3 \times 1\ mmhos/cm)$

$3X = 7623$

$X = 2541\ ft^3$ native soil required to dilute salt

Or:

dilution volume = $\dfrac{[(measured\ EC) - (target\ EC)] \times contaminated\ volume,\ ft^3}{(target\ EC - Bkgd\ EC)}$

$= \dfrac{[(25\ mmhos/cm - 4.0\ mmhos/cm)] \times 363\ ft^3}{(4.0\ mmhos/cm - 1\ mmhos/cm)}$

$= 2541\ ft^3$

Total treatment volume = contaminated volume + dilution volume

$= 363\ ft^3 + 2541\ ft^3$

$= 2904\ ft^3$

Or:

Treatment volume = $\dfrac{[(measured\ EC) - (bkgd\ EC)] \times contaminated\ volume,\ ft^3}{(target\ EC - bkgd\ EC)}$

$= \dfrac{[(25\ mmhos/cm - 1.0\ mmhos/cm] \times 363\ ft^3}{4\ mmhos/cm - 1.0\ mmhos/cm}$

$= 2904\ ft^3$

Determine the acreage required, assuming disking to a depth of 6 in.:

$2{,}904\ ft^3 \times 1/43{,}560\ ft^2/acre \times 6\ in/12\ in/ft = 0.13$ acres

6. Could the land be treated with fertilizer and gypsum? How much fertilizer and gypsum would be required? Assume the soil has a CEC of 50 meq/100g and an ESP of 59%.

Determine the acre-ft of contaminated soil:

$363\ ft^3 / 43{,}560\ ft^2/acre = 0.01$ acre-ft

Calculate the charge requirement:

Charge requirement = $\dfrac{(59\ actual\ ESP - 12\ acceptable\ ESP)}{100} \times 50\ meq/100\ g = 23.5\ meq/100\ g$

Calculate the gypsum requirement:

Gypsum requirement = 23.5 meq/100g × 1.7 tons gypsum/acre-ft of soil

= 40 tons gypsum/acre-ft × 0.01 acre-ft soil = 0.4 tons gypsum

= 0.4 tons gypsum x 2000 lb/ton

= 800 lb gypsum

Determine the weight of carbon contained in the spilled oil:

4919 lb oil × 0.78 carbon = 3837 lb carbon

Determine the amount of nitrogen required to achieve the desired C:N ratio of 150:1:

3837 lb carbon/150 lb carbon/lb N = 26 lb N required.

Determine nitrogen fertilizer requirement, using ammonium sulfate (21% N):
> 26 lb N required / 0.21 lb N/lb ammonium sulfate = 124 lb. ammonium sulfate required

Determine the P & K required to achieve a 4:1:1 N:P:K ratio:
> 26 lb N required/4 lb N/lb of P or K = 6.5 lb P and 6.5 lb K required

Determine phosphorus fertilizer requirement, using superphosphate (9.2%P):
> 6.5 lb P required/0.092 lb P/lb superphosphate = 71 lb superphosphate required

Determine the potassium fertilizer requirement, using muriate of potash (51% K):
> 6.5 lb K required / 0.51 lb. K/lb muriate of potash = 13 lb muriate of potash required

7. Describe the fertilizer and gypsum application procedure.

Apply all of the amendments in one application, because of the small amount of nitrogen required:

Apply 124 lb ammonium sulfate, 71 lb superphosphate, 13 lb muriate of potash and 0.4 tons gypsum.

Disk and cross-disk to a minimum of six in. to incorporate amendments.

10.3.3   Solution to Problem #3

A water flood storage tank is located in a wheat field. A pipeline leak caused 210 bbl of produced water to impact into the vegetated area. Sampling and analysis showed the spill area SPEC = 62.2 mmhos/cm, CEC = 25 meq/100 g, ESP = 67% and the pH = 9.3. The leak affected 0.2 acres land to a depth of 6 in.

1. What is the resulting TDS of the soil?

From Table 2.5, the soil resource is clay loam.
From Table 8.4, assume loam has saturated paste equivalent of 38% moisture.

$$38 \text{ g water}/100 \text{ g soil} = 380 \text{ g water/kg soil}$$
$$= 380 \text{ g water/kg soil} \times 1\text{ml/g} \times 1/1000 \text{ ml/L}$$
$$= 0.38 \text{ L/kg soil}$$
$$\text{TDS, mg/L} = 613 \times 62.2 = 38,129 \text{ mg/L}$$
$$\text{TDS, mg/kg} = 38,129 \text{ mg/L} \times 0.38 \text{ L/kg} = 14,489 \text{ mg/kg}$$

2. Can this site be remediated with only gypsum and sulfur?

No, some form of salt dilution will be required. If gypsum and sulfur are used in conjunction with dilution, a treatment area SPEC of 6.0 mmhos/cm can be used as the treatment standard.

3. Assuming the spill contaminated 0.2 acres of farmland to a depth of 6 in., how much gypsum and sulfur would be required to return the soil for wheat production?

Determine volume of contaminated soil:
> 0.20 acres $\times 43,560$ ft^2/acre $\times 6$ in. depth/12 in./ft = 4,356 ft^3 contaminated soil

Determine the acre-ft of contaminated soil:
> 4,356 ft^3 / 43,560 ft^2/acre = 0.10 acre-ft

Calculate the charge requirement:
$$\text{Charge requirement} = \frac{(67 \text{ actual ESP} - 12 \text{ acceptable ESP}) \times 25 \text{ meq/100 g CEC}}{100}$$
$$= 13.8 \text{ meq/100 g}$$

Half of the charge requirement should be supplied by elemental sulfur for pH adjustment, and half supplied by a calcium source such as gypsum.

Calculate the gypsum requirement:

$$\text{Gypsum requirement} = 13.8/2 \times 1.7 \text{ tons gypsum/acre-ft of soil} \times 0.10 \text{ acre-ft soil}$$
$$= 1.2 \text{ tons gypsum (CaSO}_4 \cdot 2\text{H}_2\text{O)}$$

Calculate the elemental sulfur requirement:

$$\text{Sulfur requirement} = 13.8/2 \times 0.3 \text{ tons sulfur/acre-ft of soil} \times 0.10 \text{ acre-ft soil}$$
$$= 0.2 \text{ tons sulfur (S)}$$

4. How much land is required to achieve a treatment area SP EC of 6.0 mmhos/cm, assuming the native soil has a SP EC of 1.0 mmhos/cm?

Calculate the native soil required to achieve an SPEC of 6 mmhos/cm in treatment zone:

$$6 \text{ mmhos/cm} = \frac{(4{,}356 \text{ ft}^3 \text{ contaminated} \times 62.2 \text{ mmhos/cm}) + (X \text{ ft}^3 \text{ native} \times 1 \text{ mmhos/cm})}{(4356 \text{ ft}^3 \text{ contaminated soil} + X \text{ ft}^3 \text{ native soil})}$$

Solve for X:

$$26{,}136 + 6 X \text{ ft}^3 = 270{,}943 + X \text{ ft}^3$$
$$5X = 236{,}095$$
$$X = 48{,}961 \text{ ft}^3$$

X = 48,968 ft^3 native soil required to reach an SPEC of 6.0 mmhos/cm in the treatment area

Or:

$$\text{Dilution volume} = \frac{[(\text{measured EC}) - (\text{target EC})] \times \text{contaminated volume, ft}^3}{(\text{target EC} - \text{bkgd EC})}$$

$$= \frac{[(62.2 \text{ mmhos/cm} - 6.0 \text{ mmhos/cm})] \times 4356 \text{ ft}^3}{(6.0 \text{ mmhos/cm} - 1 \text{ mmhos/cm})}$$

$$= 48{,}961 \text{ ft}^3$$

Determine the total soil volume (native soil + contaminated soil):

4356 ft^3 contaminated soil + 48,961 ft^3 native soil = 53,317 ft^3

Or:

$$\text{Treatment volume} = \frac{[(\text{measured EC}) - (\text{bkgd EC})] \times \text{contaminated volume, ft}^3}{(\text{target EC} - \text{bkgd EC})}$$

$$= \frac{[(62.2 \text{ mmhos/cm} - 1.0 \text{ mmhos/cm}] \times 4{,}356 \text{ ft}^3}{(6 \text{ mmhos/cm} - 1.0 \text{ mmhos/cm})}$$

$$= 53{,}317 \text{ ft}^3$$

Determine the acreage required, assuming a disk operation to a depth of 6 in.:

(53,317 ft^3/43,560 ft^2/acre) $\times$ 6 in./12 in./ft = 2.45 acres

5. How deep should one spread the contaminated soil?

4356 ft^3/106,722 ft^2 per 2.45 acres = 0.041 ft deep
0.041 ft $\times$ 12 in/ft = 0.5 in.

6. What steps should one take to land treat site?

1. Excavate the contaminated soil and spread over 2.45 acres to a depth 0.5 in.
2. Apply 1.2 tons gypsum and 0.2 tons elemental sulfur to the receiving area.

3. Disk and crossdisk receiving area to distribute contaminated soil and amendments to a depth of 6 in.
4. Dust the excavated area with 100 lb gypsum and back-fill with clean topsoil.
5. Treat the excavated area and receiving soil area with 50 lb balanced fertilizer (i.e., 13-13-13) to encourage revegetation.

10.3.4   Solution to Problem #4

Smith Well #4 is located at a farm near Kingsville, TX. The *reserve* pit is 100 ft ×150 ft and the drilling muds are approximately 3 ft deep.

1. Calculate the 1:1 EC (mmhos/cm):

$$SP\ EC = \frac{100 \times 1{:}1\ EC}{SP\%\ moisture}$$

$$1{:}1\ EC = \frac{SP\%\ moisture \times SP\ EC}{100}$$

$$1{:}1\ EC = \frac{98.4\% \times 62.2\ mmhos/cm}{100}$$

$$1{:}1\ EC = 61.2\ mmhos/cm$$

2. Compare the 1:1 EC to the soluble cations and anions and show your

Calculations. Explain the results. (Hint: meq/L Cl = mg/L ÷ 35.5 mg/meq.)
EC, mmhos/cm × 10 = ∑ cations, meq/L = ∑ anions, meq/L
1:1 EC × 10 = ∑ Soluble Cations (meq/L)
61.2 × 10 = ∑ Na, Ca, Mg, K (meq/L)
612 = 514.3 + 38.3 + 4.4 + 10.5
612 = 568

The calculation results are within 10%, which is acceptable. The majority of the EC is contributed by sodium (Na). Other cations such as barium and strontium were not analyzed, but generally are not major contributors to EC.

1:1 EC × 10 = ∑ soluble anions (meq/L)
Chloride is given in ppm (mg/l) and must be converted:
18,671 mg chloride/L ÷ 35.5 mg/meq = 525.9 meq Cl/L
61.2 × 10 = ∑ Cl, SO_4, HCO_3, CO_3, NO_3
612 = 525.9 + SO_4 + HCO_3 + CO_3 + NO_3
612 ≈ 525.9

Results are >10% different. This is because not all of the common anions were analyzed. Some of the EC may be due to sulfate, bicarbonate, etc. However ~ 86% of the EC is conserved as chloride. Primary salt in mud system is NaCl.

3. Which parameters exceed the regulatory thresholds under Texas regulations for landspreading?

Chlorides >3000 mg/L for landspreading on any property (see Chapter 7)

4. Are there any other parameters that should be considered to ensure the land is returned to full agronomic potential?

The land should be returned to a level at which sorghum will grow. (Hint: Look at Plate 12.)

The SPEC should not exceed 7–8 mmhos/cm to avoid significant yield decrease
SAR should be less than 12
ESP should be less than 15%
Oil & Grease < 1.0%

5a. Calculate the volume of soil needed for dilution of the limiting constituent for landspreading. You may have to calculate several parameters to determine which is the most limiting.

$$\text{Volume of wet drilling muds} = 100 \text{ ft} \times 150 \text{ ft} \times 3 \text{ ft}$$
$$= 45,000 \text{ cu ft}$$

$$\text{Volume of dry drilling muds} = \text{wet muds} \times 100 \: / \: (\% \text{ moisture} + 100)$$
$$= (45,000 \text{ ft}^3 \times 100) \: / \: (35.1 + 100)$$
$$= 33,309 \text{ ft}^3$$

10.3.4.1   EC Dilution.   Calculate soil required to dilute EC. We recommend an EC = 4 mmhos/cm as a target, because the SP moisture of the mud system is so much higher than SP moisture of receiving background soil.

$$\text{Dilution volume} = \frac{[(\text{measured EC}) - (\text{target EC})] \times \text{contaminated volume, ft}^3}{(\text{target EC} - \text{background EC})}$$

$$= \frac{[(62.2 \text{ mmhos/cm}) - (4 \text{ mmhos/cm})] \times 33,309 \text{ ft}^3}{(4 \text{ mmhos/cm} - 0.8 \text{ mmhos/cm})}$$

$$= 605,807 \text{ ft}^3$$

$$\text{Total soil volume (ft}^3) = \text{contaminated volume} + \text{dilution volume}$$
$$= 33,309 \text{ ft}^3 + 605,807 \text{ ft}^3$$
$$= 639,116 \text{ ft}^3$$

10.3.4.2   Chloride Dilution.   Calculate soil required to dilute chloride level of 3,000 ppm:

$$\text{Dilution volume} = \frac{[(\text{measured Cl}) - (\text{target Cl})] \times \text{contaminated volume, ft}^3}{(\text{target Cl} - \text{background Cl})}$$

$$= \frac{[(18,671\text{mg/L}) - (3,000 \text{ mg/L})] \times 33,309 \text{ ft}^3}{(3000 \text{ mg/L} - 46 \text{ mg/L})}$$

$$= 176,705 \text{ ft}^3$$

$$\text{Total soil volume (ft}^3) = \text{contaminated volume} + \text{dilution volume}$$
$$= 33,309 \text{ ft}^3 + 176,705 \text{ ft}^3$$
$$= 210,014 \text{ ft}^3$$

10.3.4.3   TPH Dilution.

$$\text{Dilution volume} = \frac{[(\text{measured TPH}) - (\text{target TPH})] \times \text{contaminated volume, ft}^3}{(\text{target TPH} - \text{background TPH})}$$

$$= \frac{[(1.3\ \%) - (1.0\ \%)] \times 33,309\ ft^3}{(1.0\ \% - 0.09.\ \%)}$$

$$= 11,103\ ft^3$$

Total soil volume (ft3) = contaminated volume + dilution volume
$$= 33,309\ ft^3 + 11,103\ ft^3$$
$$= 44,412\ ft^3$$

5b. Determine the amount of land necessary for landspreading.

For this example, we landspread the waste solids so the land has an SPEC value of 4 mmhos/cm. We will use a 6-in. soil layer (lift) to incorporate the waste into, the typical disc depth. If the equipment is available it is possible to blend to a deeper depth.

Acreage necessary for landspreading:

$$\frac{639,116\ ft^3}{21,780\ ft^3/acre\ 6\ in.} = 29.3\ acres$$

5c. Calculate the application depth of the drilling muds for landspreading

Volume of drilling muds $= 33,309\ ft^3$

Area of landspreading $= 29.3\ acres \times 43,560\ ft^2/acre$
$$= 1,276,308\ ft^2$$

Depth of landspreading $= 33,309\ ft^3/1,276,308\ ft^2$
$$= 0.03\ ft$$
$$= 0.03\ ft \times 12\ in./ft$$
$$= 0.4\ in.$$

To achieve this thin layer, we spread mud thickly at first over a small area, blending it into the background soil using mechanical incorporation. Then pick the larger volume up and spread over the larger area. Also, thin layer spreading can be performed using a hurricane machine.

5d. Calculate the application rates of any necessary amendments.

Because the drilling muds are diluted so much to manage salt, the TPH levels become inconsequential and do not require amendments.
Although salt has been diluted we still need to determine if sodicity parameters require amendments. Sodicity parameters are SAR and ESP.

$$\text{Calculated SAR} = \frac{(\text{waste vol.})(\text{waste SAR}) + (\text{dilution vol.})(\text{background SAR})}{\text{total soil volume}}$$

$$= \frac{(33,309\ ft^3)(111.3\ SAR) + (605,807\ ft^3)(1.6\ SAR)}{639,116\ ft^3}$$

$$= 7.3$$

$$\text{Calc. ESP} = \frac{(\text{waste vol.})(\text{waste ESP}) + (\text{dilution vol.})(\text{background ESP})}{\text{total soil volume}}$$

$$= \frac{(33,309\ ft^3)(78.9\ \%) + (605,807\ ft^3)(6.6\ \%)}{639,116\ ft^3}$$

$$= 10.4\%$$

Neither the SAR nor ESP exceeds the criteria in Appendix I. However, anytime waste drilling fluids are spread on agricultural soil, we recommend a blanket application of gypsum be applied at the rate of 500 lb/acre of treated soil. The area should also receive a minimum 100 lb/acre of 13-13-13 balanced fertilizer to improve fertility status for revegetation.

5e. Write recommendations for the landspreading application.

    a) Remove any free liquid phases from the pit.
    b) Remove pit solids and stained soil from sidewall and spread evenly over 29.3 acres receiving soil at a depth of 0.4 in.
    c) Apply 172 lb gypsum to pit floor (rate of 500 lb/acre $\times$ 15,000 ft^2/43,560 ft^2/ac)
    d) Bring in fresh top soil as necessary to restore original grade and contour
    e) Apply 14,700 lb gypsum to landspread area (500 lb/acre $\times$ 29.4 acres)
    f) Apply 2940 lb 13-13-13 fertilizer to landspread area (100 lb/acre $\times$ 29.4 acres)
    g) Disk and cross-disk waste solids and amendments into receiving soil.
    h) The area may be reseeded or allowed to revegetate naturally.

6. Calculate land requirement to spread the Smith Well #4 reserve pit solids in Oklahoma.

    SPEC drilling mud = 62.2 mmhos/cm
    Background soil EC = 0.8 mmhos/cm
    Specific gravity of mud = 91 lb/ft^3 and 12.1 lb/gal are assumed values
    Volume of waste solids = 45,000 ft^3 or 1667 yd^3

10.4   SOIL LOADING FORMULA FOR OKLAHOMA

For Oklahoma computing formula we need to convert EC saturated paste to 1:1 (by weight) EC for both drilling fluids and background soil. We calculated the 1:1 EC in step 1, yielding a value of 61.2 mmhos/cm or 61,200 µmhos/cm. Note: Oklahoma uses the conversion factor of 640 x EC to obtain the TDS value.

$$SPEC = \frac{100 \times 1{:}1\ EC}{SP\ \%\ moisture}$$

$$1{:}1\ EC = \frac{SP\ \%\ moisture \times SPEC}{100}$$

$$= 0.2\ mmhos/cm\ or\ 200\ \mu mhos/cm$$

EC of receiving soil 200 µmhos/cm $\times$ 0.64 = 128 mg/L TDS
Total Dissolved Solids in receiving soil 128 mg/L $\times$ 2 = 256 lb/acre TDS in receiving soil.
6000 lbs/acre TDS – 256 lbs/ac TDS in receiving soil = maximum TDS allowed to be applied 5744 (lb/acre).
EC of materials to be applied 61,200 µmhos/cm $\times$ 0.64 = 39,168 mg/L TDS.
Maximum TDS to be applied (5,744 lb/ac)/(39,168 mg/L $\times$ 0.000001) = maximum weight of materials to be applied 146,650 lb/acre.

10.4.1   For Solid Materials

Maximum weight to be applied $\dfrac{(146,650\ lb/acre)}{(sample\ weight\ 12.1\ lb/gal \times 202\ gal/yd^3)}$

    = Maximum loading 59.9 yd^3/acre.

Total volume of materials to applied 1,667 yd³/maximum loading 59.9 yd³/acre
= minimum acres required 27.8 acres.

10.4.2   Chlorides

Cl of receiving soil 46 mg/L × 2 = 92 lbs Cl/ac
3,500 lbs/ac Cl – 92 lbs/acre TDS in receiving soil = maximum (lb/acre) Cl to be applied 3408 lbs/acre Cl.
Maximum Cl to be applied (3,408 lb/acre) ÷ (18,671 mg/L × 0.000001) = maximum weight of materials to be applied 182,529 lb/acre.

10.4.3   For Solid Materials

Maximum weight to be applied $\dfrac{(182,529 \text{ lb/acre})}{(\text{sample weight } 12.1 \text{ lb/gal} \times 202)}$

= maximum loading 74.7 yd³/acre.

Total volume of materials to applied 1,667 yd³ / maximum loading 74.7 yd³/acre = minimum area required 22.3 acres.

10.4.4   Solution to Problem #5

Use laboratory data for Smith Well #4 *shale* pit (Table 10.1) in this problem set. Smith Well #4 is located in an elevated freshwater wetland near Baton Rouge, LA. The pit is 75 ft × 50 ft and the drilling muds and cuttings are approximately 5 ft deep.

1. What are the appropriate 29-B criteria for pit closure in elevated wetlands in Louisiana?

Parameter	Elevated Wetlands Criteria
pH	6–9
SPEC	<8 mmhos/cm
SAR	<14
ESP	<25%
Arsenic	<10 ppm
Barium	<20,000 ppm
Cadmium	<10 ppm
Chromium	<500 ppm
Lead	<500 ppm
Mercury	<10 ppm
Selenium	<10 ppm
Silver	<200 ppm
Zinc	<500 ppm
Oil & Grease	<1%

2. Calculate the 1:1 EC (mmhos/cm):

$$\text{SPEC} = \frac{100 \times 1{:}1 \text{ EC}}{\text{SP moisture}}$$

$$1{:}1 \text{ EC} = \frac{\text{SPmoisture} \times \text{SPEC}}{100}$$

1:1 EC = $\dfrac{47.7 \times 13.2}{100}$ = 6.3 mmhos/cm

3. Compare the 1:1 EC to the soluble cations and anions and show your calculations. Explain the results. (Hint: 1 meq Cl/L = 35.5 mg Cl/L.)

1:1 EC × 10 = ∑ soluble cations (meq/l)
6.3 × 10 = ∑ sol Na, Ca, Mg, K (meq/l)
63 ≈ 44.0 + 8.7 + 1.3 + 9.4 = 63.8
63 ≈ 63.8

Results are within 10%, which is acceptable. The majority of the EC is contributed by sodium.

1:1 EC × 10 = sum soluble anions (meq/L)
Chloride, the only anion analyzed, is given as 1843 mg/L and must be converted to meq/L.
1,843 mg/L Cl × 1/35.5 mg/meq = 51.9 meq Cl/L
6.3 mmhos/cm × 10 = ∑ sol anions (meq/L)
63 meq/L × 51.9 meq/L

Results are >10% different, because not all of the common anions were analyzed. Some of the EC may also be due to sulfates, carbonates or bicarbonates. We can see that 82% (51.9/63) of the EC is contributed from chlorides.

4. Which parameters exceed the regulatory thresholds under Louisiana 29-B regulations for land-spreading?

Oil and Grease >1.0%
SAR >14
SPEC >8

5. Are there any other parameters that should be considered to ensure the land is returned to its full potential? The land should be returned to its original wetland condition.
If the parameters listed above are met the land should return to its original wetland condition.

6. a) Calculate the volume of soil needed for dilution of the limiting constituent for landspreading. You may have to calculate several different items to determine which is the most limiting.

Volume of wet drilling muds = 75 ft × 50 ft × 5 ft = 18,750 ft^3

Volume of dry drilling muds = Wet muds × 100 / (% moisture + 100)

$$= 18,750 \times 100 \ / \ (27.2 + 100)$$

$$= 14,740 \ ft^3$$

Two options are available for Oil & Grease. We may either landspread it to <1.0% and walk away from the site or we may bioremediate the oil to <1.0%. There is no time limit for bioremediation, but considering we are in an elevated freshwater wetland *you must keep any oil from moving off of the site.*

10.4.4.1 O&G Dilution at 1%.   Calculate soil required to dilute to an O&G of <1.0%

Dilution volume = $\dfrac{[(\text{measured O\&G}) - (\text{desired O\&G})] \times \text{contaminated volume (cu ft)}}{(\text{desired O\&G} - \text{background O\&G})}$

Dilution volume = $\dfrac{\{(8.5\%) - (1.0\%)\} \times (14{,}740 \text{ cu ft})}{(1.0\% - 0.1\%)}$

Dilution volume = 122,833 ft^3

Total soil volume (ft^3) = contaminated volume + dilution volume

$$= 14{,}740 \text{ ft}^3 + 122{,}833 \text{ ft}^3$$

Total soil volume (ft^3) = 137,573 ft^3

10.4.4.2   SAR Dilution.   Calculate soil required to dilute to a SAR of 14.

Dilution volume = $\dfrac{\{(\text{measured SAR}) - (\text{desired SAR})\} \times \text{contaminated volume (ft}^3)}{(\text{desired SAR} - \text{background SAR})}$

$$= \dfrac{\{(19.7 \text{ mmhos/cm}) - (14 \text{ mmhos/cm})\} \times (14{,}740 \text{ ft}^3)}{(14 \text{ mmhos/cm} - 1.6 \text{ mmhos/cm})}$$

Dilution volume = 6,776 ft^3

Total soil volume (ft^3) = contaminated volume + dilution volume
$$= 14{,}740 \text{ ft}^3 + 6{,}776 \text{ ft}^3$$

Total soil volume = 21,516 ft^3

10.4.4.3   EC Dilution.   Calculate soil required to dilute to a SPEC of 8 mmhos/cm

Dilution volume = $\dfrac{[(\text{measured EC}) - (\text{desired EC})] \times \text{contaminated volume (ft}^3)}{(\text{desired EC} - \text{background EC})}$

$$= \dfrac{\{(13.2 \text{ mmhos/cm}) - (8 \text{ mmhos/cm})\} \times (14{,}740 \text{ ft}^3)}{(8 \text{ mmhos/cm} - 0.8 \text{ mmhos/cm})}$$

Dilution volume = 10,646 ft^3

Total soil volume (ft^3) = contaminated volume + dilution volume
$$= 14.740 \text{ ft}^3 + 10{,}646 \text{ ft}^3$$

Total soil volume = 25,386 ft^3

O&G is the limiting constituent and requires 137,573 ft^3 for landspreading. If we want to bioremediate the oil the SPEC of 8 mmhos/cm is the limiting factor and the total volume is 25,386 ft^3. Calculate the O&G loading rate for a SPEC of 8 mmhos/cm

Calc. O&G = $\dfrac{(\text{contaminated vol})(\text{measured O\&G}) + (\text{dilution vol.})(\text{background O\&G})}{\text{total soil volume}}$

$$= \dfrac{(14{,}740 \text{ ft}^3 \times 8.5 \ \%) + (10{,}646 \text{ ft}^3)(0.1\%)}{25{,}386 \text{ ft}^3}$$

Calc. O&G = 5.4%

Since there is no time limit for remediation it is possible to remediate down to 1%. But an O&G at 5.4% can inhibit plant growth. If it is an environmentally sensitive area, then we recommend a lower O&G loading rate, at 2.0% or less.

b) Determine the amount of land necessary for landspreading.

We desire to landspread the waste so the resulting land has an SPEC value of 8.0 mmhos/cm and O&G level of 5.4%. We will use 6 in. as the application depth, since most farm equipment is suited to plowing to a depth of 6 in. For O&G remediation it is recommended to stay in the surface 6 in. to achieve proper aeration.

Acreage necessary for landspreading

$$\frac{25,386 \text{ cu ft}}{21,780 \text{ cu ft/acre-6 in.}} = 1.17 \text{ acre-6 in.}$$

c) Calculate the application depth of the drilling muds for landspreading.

Volume of drilling muds = 14,740 cu ft

Area of landspreading = 1.17 acres × 43,560 sq ft/acre
= 50,965 sq ft

Depth of landspreading = 14,740 cu ft/50,965 sq ft

Depth of landspreading = 0.29 ft = 3.5 in.

Landspread the drilling muds and cuttings a depth of 3.5 in. over 1.17 acres.

d) Calculate the application rates of any necessary amendments.

Since the salts have been diluted to 8.0 mmhos/cm SPEC and the SAR and ESP are below regulatory limits there is no need for gypsum or sulfur amendments.

To remediate the oil, fertilizer must be added. The N:P:K calculations are as follows:

carbon content in mg/kg:

C, %= O&G% × 0.78
C, %= 5.4 × 0.78
C, %= 4.2
C, mg/kg= 4.2% × 10,000
C, mg/kg= 42,000 mg/kg

Nitrogen requirement:

C/N= 150/1
(42,000 mg/kg)/N= 150/1
N= 280 mg/kg
N= 280 mg/kg × 2
N= 560 lb/acre-6 in.

Treatment area occupies 1.17 acres

N = 560 lb/acre-6 in. × 1.17 acre
N = 655 lb

NPK 4:1:1

P,K = 655/4
P,K = 164 lb

The above calculations are for the quantity of N:P:K to add. The actual amount of fertilizer depends upon the %age of each element in the fertilizer. Ammonium nitrate, concentrated superphosphate and muriate of potash are common amendment.

NOTE: Be careful when adding fertilizer marked as #-#-# such as 12-10-8. The first number, 12, stands for% N. The number 10 stands for percent P_2O_5, not percent P. The number 8 stands for the percent K_2O, not percent K.

Pounds ammonium nitrate (33% N)

ammonium nitrate = 655 / 0.33
ammonium nitrate = 1985 lb

Pounds concentrated superphosphate (21% P)

Concentrated superphosphate = 164/0.21
Concentrated superphosphate = 781 lb

Pounds muriate of potash (51% K)

Muriate of potash = 164/0.51
= 322 lb

e) Write recommendations for the landspreading application.

1. Remove any free liquid phases from the pit sections.
2. Remove pit solids and sludge from pit and sidewalls and spread evenly over 1.17 acres a depth of 3.5 in.
3. Push pit levees into cover pit residues and restore pit area to the original contour. The pit floor should have a minimum 5 ft soil cover.
4. Add half of the ammonium nitrate, 993 lb (33% N), all of the superphosphate, 781 lb (21% P) and all of the muriate of potash, 322 lb (51% K), to the spill and receiving areas.
5. Disk and cross-disk to 6 in. to distribute fertilizer and aerate treatment zone.
6. After 6– 8 weeks add half of the remaining ammonium nitrate (496 lb), disk and cross-disk to distribute fertilizer and aerate treatment zone.
7. After 6–8 weeks add final 496 lb ammonium nitrate, disk and cross-disk to distribute fertilizer and aerate treatment zone.
8. Irrigate as necessary to maintain treatment zone between 50 and 80% of field capacity, for example, 1 in. water/week.
10. After 6–8 weeks sample site to demonstrate cleanup level <1% O&G. Retreat remediation, as above using the new O&G concentration if necessary.

10.4.5   Solution to Problem #6

Use laboratory data for Smith Well #4 shale pit, in this problem set.

Smith Well #4 is located on a farm near Tulsa, OK. The pit is 75 ft × 50 ft and the drilling muds are approximately 5 ft deep.

1. Calculate the wet weight of pit solids in lb/gal.

The wet weight of the pit solids is equivalent to the weight of the dry solids plus the weight of the water contained in the solids.
Calculate the weight of water in the solids:
The solids contain 27.2% water. This water is held in the soil pores, so it does not contribute to the soil volume, only to the bulk soil weight.
100 grams of these solids contain 27.2 ml of water. This is equivalent to 27.2 g of water (density of water = 1.0 g/mL).

Calculate the weight of dry soil in the solids:
For these solids, the CEC = 8.0 meq/100g, indicating a sandy loam material.
Dry bulk density of sandy loams generally range from 1.3 to 1.6 g/cm^3, so assume a bulk density (dry) of 1.5 g/cm^3 for this case.
100 g of dry soil/1.5 g/cm^3 = 66.7 cm^3 = 66.7 mL
Wet solids density = (100 g soil + 27.2 g water) / 66.7 mL solids = 1.91 g/mL.

Convert the density to the required units:

1. 91 g/mL × 1000 mL/L × 3.8 L/gal × 1 lb/454g = 16.0 lb/gal

2. Calculate the dry weight% of the solids.

$$\frac{100 \text{ g dry solids} \times 100}{(100 \text{ g dry solids} + 27.2 \text{ ml water})} = 78.6\% \text{ dry solids}$$

3. Calculate the volume of solids in the pit in barrels.

   Calculate the pit solids volume in cubic feet:

   75 ft × 50 ft × 5 ft = 18,750 ft^3 solids

   Convert volume from ft^3 to barrels:

   42 gal/bbl × 3.785 L/gal × 1000 cm^3/L = 158,970 cm^3/bbl

   158,970 cm^3/bbl × 1 in^3/16.39 cm^3 × 1 ft^3/1,728 in^3 = 5.61 ft^3/bbl

   18,750 ft^3 solids/ 5.61 ft^3/bbl = 3342 bbl solids

4. Using Oklahoma Corporation Commission calculations (OAC 165:10 Appendix 1 of rule), determine the acreage required for landspreading of solids using the dry weight application limitation (200,000 lb/acre dry solids).

   Wet weight of solids 16.0 lb/gal × 0.79 % dry weight = 12.64 lb/gal dry weight

   12.64 lb/gal dry weight × 42 gal/bbl = 531 lb/bbl

   200,000 lb/acre dry weight / 531 lb/bbl = maximum 377 bbl/acre

   Total volume to be applied 3342 bbl/maximum bbl/acre 377 = minimum 8.9 acres required

5. Calculate the ppm TDS in the receiving soil (Bkgd).

   The saturated paste EC of the receiving soil is 0.8 mmhos/cm.

   0.8 mmhos/cm × 640 (mg salt/L water/ mmhos/cm) = 512 mg salt/L water.

   This soil has a saturated paste moisture percent of 27.2%, meaning 100 g of soil holds 27.2 ml of water, which is equivalent to 0.272 L water/kg soil.

   512 mg salt/L water × 0.272 L water/kg soil = 139 mg salt/kg soil (ppm by wt)

6. Calculate the mg/kg TDS of the pit solids.

   The saturated paste EC of the pit solids is 13.2 mmhos/cm.

   13.2 mmhos/cm × 640 (mg salt/L water/ mmhos/cm) = 8,448 mg salt/L water

The solids have a saturated paste moisture of 47.7%, meaning 100 g of soil holds 47.7 ml of water, which is equivalent to 0.477 L water/kg soil

8448 mg salt/L water × 0.477 L water/kg soil = 4030 mg salt/kg soil (ppm by wt)

7. Using OCC calculations (OAC 165:10 Appendix 1 of rule), determine the acreage required for landspreading of solids using the total dissolved solids application limitation (6000 lb. TDS/acre).

   139 mg/L TDS in soil × 2 = 278 lb/acre TDS in receiving soil

   6000 lb./ac TDS – 278 lb/ac TDS in receiving soil = maximum TDS to be applied 5,722

   Maximum TDS lb/acre to be applied 5722 / (4,030 mg/kg TDS of pit solids × 0.000001) = Maximum materials to be applied 1,419,850 lb/acre

   Using the dry solids density calculated in Item #4, above:

   Maximum lb/ac 1,419,850/(12.64 lb/gal dry weight × 42 gal/bbl) = Maximum 2674 bbl/ac

   Using the total volume of solids calculated in Item #3, above:
   Total volume to be applied 3324/maximum bbl/ac to be applied 2,674 = minimum 1.24 acres required.

8. Calculate the mg/kg Oil & Grease in the pit solids

   The Oil & Grease percentage of the solids is 8.5%, so 100 g of solids contain 8.5 g of oil and grease (85 g/kg).

   (85 g oil & grease/ kg solids) × 1000 mg/g = 85,000 mg/kg solids (ppm)

9. Using OCC calculations (OAC 165:10 Appendix 1 of rule), determine the acreage required for landspreading of solids using the oil & grease (O&G) application limitation (40,000 lb O&G/acre).

   40,000 lb/acre O&G / (85,000 mg/kg O&G in solids × 0.000001) = maximum to be applied 470,588 lb/acre of solids

   Using the dry solids density calculated in Item #4, above:

   Maximum lb/acre of solids to be applied 470,588/(12.64 lb/gal dry wt × 42 gal/bbl) = maximum 886 bbl/ac

   Using the total volume of solids calculated in item #3, above:
   Total volume to be applied 3324/maximum bbl/ac to be applied 886 = minimum 3.75 acres required.

10. Based upon the OCC regulations for land application of solids, what is the minimum land requirement for disposal of these pit solids?

    Minimum land requirement for solids disposal, using the dry solids application threshold of 200,000 lb/acre = 8.8 acres (see item #4, above)

    Minimum land requirement for solids disposal, using the TDS threshold of 6000 lb/acre = 1.24 acres (see item #7, above)

    Minimum land requirement for solids disposal, using the Oil & Grease threshold of 40,000 lb/acre = 3.75 acres (see item #9, above)

    The minimum land requirement is equivalent to the largest of the minimum land requirements, or 8.8 acres. Disposal of pit solids using 8.8 acres will meet all land disposal requirements.

APPENDIX I
LIMITING CONSTITUENTS FOR LAND DISPOSAL OF PETROLEUM EXTRACTION INDUSTRY WASTES

I-1 INTRODUCTION

Onshore Exploration and Production [E&P] functions generate large quantities of non-hazardous and exempt wastes. The vast majority of this volume is produced water [estimated at 98 percent], which, except for small quantities discharged into near shore marine waters or used for wildlife watering, are injected into Class II wells [40 CFR Part 146]. The remaining volume consists of spent drilling fluids, treating fluids and associated wastes. Associated wastes include, produced sand, tank bottoms and other materials circulated from wells or separated from oil, water and gas flowlines or tank battery facilities prior to custody transfer. E&P wastes typically are disposed onsite. However, a small quantity goes to offsite facilities.

This appendix develops limiting concentrations of petroleum hydrocarbons and salts allowable for one time landspreading, burial and roadspreading of E&P wastes. Parameters providing evidence of proper saline waste management include:

- Electrical Conductivity [EC]
- Sodium Adsorption Ratio [SAR]
- Exchangeable Sodium Percentage [ESP] and for hydrocarbons:
- Total Petroleum Hydrocarbons [TPH]

The limiting values for land application of E&P wastes apply to soil/waste mixtures. We recommend the following values for various use conditions.

TABLE A-I-1
LIMITING CRITERIA FOR LAND TREATMENT

		---------------------Wetlands---------------------			--------------Uplands--------------	
	Brackish	Freshwater Submerged	Freshwater Elevated	Forest/ Recreational	Farming/ Residential
pH	6–9	6–9	6–9	6–8	6–8
EC (mmhos/cm)	---	---	8–50	4–8	2–4

127

128 Appendix I

TABLE A-I-1
(Continued)

	Brackish	Freshwater Submerged	Freshwater Elevated	Forest/ Recreational	Farming/ Residential
SAR	---	---	14–100	<14	<12
CEC (meq/100g)	>15	>15	>15	>15	>15
ESP (%)	---	---	<25	<15	<15

Heavy Metals (mg/kg)	Wetlands	Uplands
Arsenic	<10	<40
Barium	<20,000	<40,000
Cadmium	<10	<10
Chromium	<500	<500
Lead	<500	<500
Mercury	<10	<10
Selenium	<10	<10
Silver	<10	<10
Zinc	<500	<500

Organics (%)	Wetlands	Uplands
Oil & Grease	<1	<2

One-time waste application producing soil/waste mixtures having values less than the above limits result in minimal impact on soil and vegetation. Except for the most sensitive crops, the waste application results in less than 15 percent reduced yield during the first year after application. Typically, total recovery of crop yield potential occurs within 3 years. However, in some cases, soil amendments such as gypsum and nitrogen containing fertilizers may be required.

I-2 BACKGROUND

I-2.1 Literature Review

Investigators [Miller and Honarvar, 1975; Ferrante, 1981; Freeman and Deuel, 1984; Nelson, 1984] identified salts and hydrocarbons, found in E&P wastes, as responsible for reduced crop yields. These contaminants, in excess of given thresholds values, result in phototoxicity. Additionally, sodium salts deteriorate soil structure [porosity and permeability] resulting in reduced plant water availability, excess water runoff and erosion. Also, salts and hydrocarbons can impact adversely surface and ground waters [Henderson, 1982; Murphy and Kehew, 1984].

I-2.2 Salinity

Soil retains moisture which dissolves some soil constituents. The dissolved material enters the soil solution as cations and anions in proportion to their solubility in the matrix. Major ions observed in E&P wastes include calcium [Ca], sodium [Na], magnesium [Mg], potassium [K], chloride [Cl], sulfate [SO_4], bicarbonate [HCO_3], carbonate [CO_3] and hydroxide [OH]. The more ions dissolved in the aqueous phase, the higher is the electrical conductivity of that phase. Accordingly, electrical conductivity [EC] becomes a measure of the ionic strength of the soil solution. Sodium Adsorption Ratio [SAR] and Exchangeable Sodium Percentage [ESP] measure the influence specific soluble ions [Na, Ca, Mg] impose on the soil. Accordingly, when viewed together, EC, SAR, & ESP, represent a measure of the available cations and ions in the soil solution. This combination is given the general term of salinity.

I-2.3   Hydrocarbons

Hydrocarbons, when released to surface soils, penetrate to varying depths depending on soil type. Oil has an affinity for clay. Accordingly, oil does not penetrate deeply into clay soils. On the other hand, oil penetrates deeply into sandy soil.

Several measures of hydrocarbons are available. EPA measures hydrocarbons by a standardized analytical protocol known as Oil and Grease [40 CFR Part 136]. This test, when applied to soils, is time consuming and costly. Since Oil and Grease [O&G] method represents the official EPA method, EPA, SW-846, Method 3540], Annex A, we use it in this appendix. However, based on extensive research [Deuel and Holliday, 1993], we recommend use of the analytical method known as Total Petroleum Hydrocarbon [TPH]. Two variations of TPH are available, TPH-Infrared and TPH-GC-FID.

I-2.3.1   TPH-Infrared.   This method [EPA 600/4-79-020, Method 418.1] provides good results on fresh spill of hydrocarbons to soils. However, for mature spills, Deuel and Holliday [1993] find the method demonstrates little-to-no decrease in hydrocarbon content as the result of bioremediation because the method reports biomass as TPH. This results in demonstrating no reduction of TPH as the result of treatment for a lag period of 1 to 2 years if co-metabolites have been used.

I-2.3.2   TPH-GC-FID.   TPH can be determined by Gas Chromatography [GC] and Flame Ionization Detector [FID]. Deuel and Holliday [1993] confirmed by Loss on Ignition the GC-FID analytical method provided a reliable method of monitoring decrease in soil oil content as treatment progresses.

I-3   DISSOLVED CONSTITUENT RELATIONSHIPS

The soluble constituents [ions] present in soil solution conduct electrical current. The conductivity depends on type and concentration of ions. This ability to conduct electrical current is electrical conductivity [EC]. EC, measured direct in water extracts of soil and solid waste (Annex B), is reported in millimhos per centimeter [mmhos/cm] which is recognized as the reciprocal of units of resistance. Dissolved salts represent the predominate ions within the soil solution. Accordingly, EC is used as an indirect measure of Total Dissolved Solids [TDS].

Barrow [1966] showed an exact correlation exists between EC and a specific salt dissolved in distilled water at low concentrations. In high concentrations, mixed salt species solutions or the presence of non-ionic dissolved components, the correlation becomes inaccurate. A statistical correlation

$$TDS = A * EC \qquad\qquad (A\text{-}I\text{-}1)$$

was developed by regression analysis. Hem [1985] found the range of "A" [slope] varied from 540 to 960 cm.mg/mmhos liter for various concentrations of dissolved ions in the presence of impurities. U.S. Salinity Laboratory staff [1954] recommended A = 640 for naturally occurring saline-sodic soils. We find for E&P soil/waste mixtures A = 613 provides a better correlation. This is based on API [1987] and EPA [1987] data review. Using the following values in Equation A-I-1:

$$EC = 4 \text{ mmhos/cm}$$
$$A = 613 \text{ cm.mg/mmhos liter}$$
$$TDS = 4 * 613$$
$$TDS = 2452 \text{ mg/liter}$$

Hydrocarbons and fine clay particles interfere with the TDS analysis [Annex C]. Accordingly, TDS is not accurate measure of soil salinity. Therefore, we recommend determining the degree of salinity of the soil using Equation A-I-1.

I-4   DISSOLVED CONSTITUENT IMPACTS

Dissolved contaminants [hydrocarbons and ions] adversely impact plants and soils.

I-4.1   Plant Impacts

Hydrocarbons and salts dissolved in the soil moisture increase the osmotic pressure of the soil solution. Plant roots in contact with this soil solution do not posses the ability to overcome the osmotic pressure of the soil and die from lack of water [Haywood and Wadleigh, 1949; U.S. Salinity Laboratory staff, 1954]. The point of permanent wilting occurs when plants cannot recover in the presence of non-saline water. Interestingly, a statistical correlation exists between soil osmotic pressure and EC:

$$\text{Osmotic Pressure [OP], atm} = 0.36 * EC \tag{A-I-2}$$

The uptake of salts from contaminated soil disrupts nutrient uptake and utilization by antagonism [Kramer, 1969]. Accordingly, excess salt alters the composition of soil solution, reducing the availability of essential plant nutrients. The result is lowered yield and crop quality. However, no one threshold salinity level exists for all plants [Maas and Hoffman, 1979]. Maas [1986] shows vegetables are more sensitive to salt than grasses and grains, Plates 12, 13 and 14. Maas [1986] developed the data for Plates 12, 13 and 14 from agricultural projects receiving salt containing water during a long period of time and could elevate the response for single time applications of waste to soils. Lunin [1967] suggests doubling the threshold limits associated with continual use system for single application situations.

The U.S. Salinity Laboratory staff [1954] developed a general crop response to salt, Table A-I-2. Basically, the Table suggest no effect on crop yield at or below EC = 2 and gross crop yield reduction or complete failure for all but the most salt tolerant species at EC = 16 or greater.

If drainage is present, excess salinity can be leached from the soil by rainfall or application of irrigation water. Also, growing salt tolerant plants aids in soil recovery since the root growth allows air and water to travel through the soil [Foth and Turk, 1972]. Salt-contaminated soil typically responds to application of calcium sulfate (gypsum). The response mechanism consists of replacing the exchangeable sodium ions with calcium ions [Oster and Rhoades, 1984]. The presence of iron in salt contaminated soil apparently aids calcium/sodium exchange [Maas, 1986]. Accordingly, for gypsiferous soils, plants tolerate EC values about 2 mmhos/cm greater than shown in Plates 12, 13 and 14.

TABLE A-I-2
GENERAL CROP RESPONSE AS A FUNCTION OF EC
(AFTER U. S. SALINITY LABORATORY STAFF, 1954)

EC, mmhos/cm	Affect on Crop Yield
0–2	None
2–4	Slight
4–8	Many crops affected
8–16	Only tolerant crops yield well
>16	Only very tolerant crops yield well

Producing or swabbing into the reserve pit increase spent mud salinity. Miller and Pesaran [1980] observed decreased plant growth in a 1:1 mud/soil mixture when the mud contained elevated salt concentrations. Miller and Pesaran [1980] test results showed for mud/soil mixtures having EC<8:

- 7 percent decrease in growth for green beans
- 13 percent decrease in growth for corn

Nelson [1984] showed an average decrease in yield of 20 percent and 38 percent for swiss chard and rye grass, respectively, when EC ranged from 6.3 to 18.6 mmhos/cm. These results demonstrate the impact high EC values.

On the other hand, Tucker [1985] reported no adverse yield impact on Bermuda grass and alfalfa at EC values of 1.3 to 5.3 and 1.7 mmhos/cm, respectively. Tucker [1985], also, reported a significant decrease in soil EC values as a function of time following soil/waste mixing. He attributed this to leaching of the salts from the plant root zone.

For sensitive areas, wetlands, working to a soil/waste mixture EC <4 is expected to result in less than a 15 percent decrease in yield for most crops. In reality, since the adverse impact of EC decreases rapidly as salts leach from the soil and, most oil and gas fields are not in sensitive areas or used to raise vegetable crops, working to an EC <8 normally is adequate. Refer to Table A-I-1 for specific guidance.

I-5 LEACHING IMPACTS

In areas of net surface infiltration, soluble salts migrate from the surface to lower soil horizons. Leaching of salts from saturated brine mud [EC >200 mmhos/cm] threatens ground water under the pit [Murphy and Kehew, 1984]. However, as previously discussed, EC >200 exceeds recommended soil/waste mixture criteria, Table A-I-1.

Further, water and associated dissolved constituents move through the soil by natural redistribution governed by pore dynamics, dispersion and diffusion. This process is similar to the method by which extracts move through a long capillary associated with a chromatograph.

Owens et al. [1985] and Bruce et al. [1985] demonstrate this phenomenon. Both studies observed redistribution of surface applied bromine [Br] by rain infiltration and percolation. Owens et al. [1985] showed more than a 7-fold decrease in bromide after percolating water through 2.4 meters [94 inches] of well-drained silt loam. Assuming the same soil conditions, a surface loading of salt equivalent to a soil/salt mixture of equal to EC = 8 mmhos/cm [TDS = 4904] results in an EC = 0.7 mmhos/cm or EC <1.2 mmhos/cm, respectively at 2.4 m [94 inches]. Additionally, assuming sodium chloride salt, the chloride concentration at 2.4 m [94 inches] corresponds to TDS <213 mg/liter and 426 mg/liter, respectively. Further, Bruce et al. [1985] demonstrated redistribution of surface bromide [Br] as high as 1800 mg/l resulted in concentrations of <20 mg/liter at 3 meters [118 inches] after nearly 4 years of leaching and 4.7 meters [46 inches/year] of rainfall. Bromide concentration decreased to 100 mg/liter at 1.5 meters and to less than detection limit at 3.8 meters [12 feet]. We consider bromide and chloride identically acting materials in soil. Accordingly, at EC - 4 or 8 mmhos/cm, chloride concentrations in the percolating water will be less than the Secondary Drinking Water Quality standard of 250 mg/liter [40 CFR §143.3] in less than 3 feet from the surface source. Thus, the selection of EC = 4 or 8 depends on land use considerations, Table-A-I-1.

I-6 SOIL CHEMISTRY

Based on mass action principles of cation exchange in soils, one expects the percentage of sodium on the exchange complex to be a function of the relative concentrations of sodium and competing cations in soil solution. The same principles hold true for E&P waste solids. The degree of sodium saturation in soils and E&P waste solids is determined from three interdependent soil chemical properties. These are:

- Cation Exchange Capacity [CEC]
- Sodium Adsorption Ratio [SAR]
- Exchangeable Sodium Percentage [ESP]

I-7   CATION EXCHANGE CAPACITY

Cation Exchange Capacity [CEC] measures the quantity of cations reversibly adsorbed per unit weight of soil. Typically, CEC is expressed in milliequivalents per 100 grams of mass [meq/100 g]. The equivalent standard metric unit is centimoles per kilogram (cmole/kg). The predominant cations held on the particle surface of soils and earthen waste solids are calcium [Ca], magnesium [Mg], sodium [Na] and potassium [K], known as the base cations. The percentage of the CEC occupied by the basic cations is called the base saturation. Fertile soils possess base saturations greater than 80 percent, consisting predominately of calcium and magnesium cations.

CEC determination requires laboratory analysis [Annex D] of field samples [Annex E]. A direct relationship exists between CEC and Exchangeable Sodium Percentage [ESP]:

$$ESP = [NaX/CEC] * 100 \qquad\qquad (A\text{-}I\text{-}3)$$

where NaX [exchangeable sodium] [Annex F] and CEC are expressed in meq/100g. ESP is calculated by an imperial relationship dependent on Sodium Adsorption Ratio [SAR] as shown in the next section.

I-8   SODIUM ADSORPTION RATIO

Good soil structure and fertility require large amounts of calcium and magnesium. These beneficial cations form salt complexes of low solubility within soils. On the other hand, sodium cations form soluble salts which dominate soil solutions resulting in deleterious impacts. Accordingly, we require a measure of impact of sodium, i.e., Sodium Adsorption Ratio [SAR].

The U.S. Department of Agriculture Salinity Laboratory [1954] developed SAR [Annex G] as an imperial index of detrimental sodium effects on soils.

$$SAR - Na/[(Ca + Mg)/2]^{\frac{1}{2}} \qquad\qquad (A\text{-}I\text{-}4)$$

where the cation values from direct chemical analysis of waste liquids or aqueous extracts of waste solids or soils are expressed in meq/liter [Annex H]. To convert concentrations from mg/liter to meq/liter simply divide the former by the milliequivalent weight of a specific cation (see 4.1 Annex G). The U.S. Salinity Laboratory staff [1954] developed an empirical *equilibrium* expression relating SAR and ESP.

$$ESP = \frac{1.475 \times SAR}{1 + 0.0147 \times SAR} \qquad\qquad (A\text{-}I\text{-}5)$$

I-9   EXCHANGEABLE SODIUM PERCENTAGE

Knowing the Sodium Adsorption Ratio [Equation A-I-4] the ESP under equilibrium conditions can be determined from Equation A-I-5. Some inaccuracy results when using formula A-I-5 on fresh releases since equilibrium is not achieved. However, the error is not significant. Accordingly, we recommend using the relationship.

I-10   SOIL CHEMISTRY IMPACTS

SAR values greater than 12 indicate high sodium concentrations in soil. This causes Ca and Mg deficiency in plants because of antagonistic reactions and reduction of solubilities [Kramer, 1969; U.S. Salinity Laboratory staff, 1954].

Sodic describes soil for which ESP>15 percent [U.S. Salinity Laboratory staff, 1954]. Soils exposed to solutions having high salt concentrations so the soil extract SAR >12.8 and ESP >15 percent risk becoming sodic. The distinguishing feature of sodic soil is lack of structure [tilth and permeability] and tendency of the soil to disperse in water. A dispersed soil devastates plants by reducing air and water supplies to the root system [Reeve and Fireman, 1967; Bresler et al. 1983].

Tucker [1985] demonstrated spent drilling fluid amended soils having SAR <15 and ESP <15 percent are required for normal plant growth and acceptable soil structure. SAR of soils can be altered easily by soil amendments [gypsum, calcium, nitrate, etc.]. Accordingly, SAR is less critical than ESP. Deuel and Brown [1980] showed the detrimental impact on soil when exposed to water having an EC = 2.8 mmhos/cm and SAR = 16.1 was directly proportioned to the calcium in the soil. The occurrence of naturally occurring or amendment excess gypsum in the soil permits disposal of sodic E&P wastes, if the ionic strength of the salt is low. Freeman and Deuel [1984] demonstrated successful pit closures [SAR <15 and ESP <15 percent] by land incorporation of E&P waste solids having SAR's >200 and ESP's >90 when salt levels in the waste solids were <4 EC, mmhos/cm. Incorporation consisted of blending waste solids with background soils at defined mixture ratios in the presence of gypsum and fertilizer.

I-11   SOIL CHEMISTRY LIMITING VALUES

Based on the above discussion, we recommend limiting the SAR to less than 12 and ESP to less than 15 percent, Table A-I-1. The Department of Agriculture [U.S. Salinity Laboratory, 1954] and various investigators accept these soil chemistry limits as a means of preventing soil sodicity. However, note the recommended values apply to the waste amended soil and not to the waste per se.

I-12   HYDROCARBON IMPACTS

E&P wastes contain crude oil, diesel fuel and lubricating oil [Miller, et al. 1980; Thoresen and Hinds, 1983; Whitfill and Boyd, 1987]. Typically, crude oil containing wastes result from well pulling operations, pipeline leaks or leaks or overflows. Also, during drilling operations, operators add crude oil or diesel lubricate the bit or decrease water loss from water base muds. Additionally, lubricating oil drips from drilling and producing machinery while in operation.

Hydrocarbon concentrations in water base muds seldom exceed 4 percent by weight [Freeman and Deuel, 1984]. Tank bottoms, emulsions and oil contaminated soil typically contain high concentrations of hydrocarbons.

Crude oil and diesel fuel consist of complex saturated and aromatic hydrocarbons [Thoresen and Hinds, 1983; Oudot et al., 1989]. Being a refined product, lubricating oil is a saturated viscose hydrocarbon containing small quantities of additives. Brown, et al. [1983] shows these hydrocarbons easily partitioned from water by solvent using separator funnels or extracted from solids by using a Soxhlet apparatus. Hydrocarbons extracted are assayed gravimetrically and reported collectively as oil and grease. The EPA protocol requires Freon 112 as the extractant, [SW-846, Method 3540]. However, for solids, the solvent of choice is methylene chloride because it efficiently extracts petroleum hydrocarbons without co-existing significant quantities of naturally occurring organic matter [Brown and Deuel, 1983].

As discussed before, Total Petroleum Hydrocarbon by Gas Chromatography with Flame Ionization Detector [TPH-GC-FID] is an excellent substitute for Oil and Grease. Typically, we use TPH in place of Oil and Grease.

Extensive research has occurred regarding the impacts of crude oil and diesel fuel on plants and soils [Schollenberger, 1968; Harper, 1939; Plice 1948; Schevendinger, 1968; Garner, 1971; Odu, 1972; Miller and Honarvar, 1975 and Miller, et al. 1980].

Phototoxic compounds consist of low molecular weight aromatic hydrocarbons present initially [diesel fuel] or form as metabolites of various degradation processes. [Plice [1948], Baker [1970], Patrick [1971], Honarvar [1975], Udo and Fayemi [1975], and Thoresen and Hinds [1983] report germination inhibition and yield reduction of row crops planted in soil containing more than 2 percent crude oil by weight. Yields decreased 50 percent from control area yields at 2 percent oil. Pal and Overcash [1978] reported reduction in growth of vegetables and row crops with oil contaminated soil containing 1 percent by weight. Bulman and Scroggins [1988] demonstrated satisfactory plant growth at 3.5 percent oil by weight but poor plant growth at oil contents exceeding 5 percent. They observed, at another site, reduced plant growth during the first season in soil contaminated with 1 and 2 percent. However, they report enhanced crop growth at 0.5 percent oil. Frankenberger and Johanson [1982] reported heavy ends of crude oil and refined products added to soil at 20 to 60 percent by weight disrupt oxidative and soil microflora activities requisite to biological remediation. These observations followed oil spills to soil.

Miller et al. [1980] observed 1 percent by weight soil loading with diesel fuel resulted in 49 to 69 percent decrease in yield of beans and corn, respectively. However, replanting after 4 months resulted in normal growth. Younkin and Johnson [1980] planted reed canarygrass in soil contaminated with 0.45 percent diesel fuel. The resulting germination decreased 69 percent and yield decreased 79 percent. During the harvest, Younkin and Johnson [1985] observed no decrease in yield [75 days after addition of diesel to the soil]. Overcash and Pal [1979] determined 1 percent oil by weight as the threshold for reduced crop yield. Concentrations of 1.5 to 2 percent by weight reduced yield by greater than 50 percent. These effects occur immediately after oil application and before the oil is biodegraded or evaporated. Further, Overcash and Pal [1979] showed tolerance to oil is a function of crop species, Table A-I-3.

These studies indicate that under hydrocarbon loadings > 1% oil & grease, E&P wastes may be detrimental toward plant growth. However, at 1% or less of mixed hydrocarbons, little or no yield reduction is expected based on existing information. Also, recovery of a site adversely impacted at a 1% loading rate is expected after a few months but not exceeding one growing season, following a one-time application.

Carr [1919] suggested oil damage resulted from reduced aeration-water interaction rather than oil induced toxicity. On the other hand, Ellis and Adams [1961] suggest iron and manganese release under low oxygen conditions contributes to phototoxicity of soil contaminated with petroleum hydrocarbons. Phototoxicity appear to reduce after assimilation of the oil by the soil.

TABLE A-I-3
OIL TOLERANCE FOR SELECTED CROPS

Crop Type	Single Oil Application
yams, carrots, rape, lawngrasses, sugarbeets	< 0.5% of soil weight
ryegrass, oat, barley, corn, wheat, beans, soybeans, tomato	< 1.5% of soil weight
red clover, peas, cotton, potato, sorghum	< 3.0% of soil weight
perennial grasses, coastal bermudagrass, trees, plantain	> 3.0% of soil weight

We conclude from these studies, E&P wastes mixed at 1 percent or less with soil presents little or no adverse effect plant yield. Further, recovery of a site contaminated at a loading rate of more than 1 percent occurs within a few months of the oil application.

I-13   OIL MOBILITY

Several investigators report low mobility of hydrocarbons in soil, particularly in clay soils. Plice [1948] observed natural segregation when liquid oil enters soil. The heavy ends [more viscous fraction] remain near the surface while the light ends penetrate more deeply. This means the more soluble fractions increase with depth [Duffey et al., 1977; Weldon [1978].

Benzene, toluene, ethylbenzene and xylenes [BETX] represent the more mobile hydrocarbons in soil [Roy and Griffin, 1985]. EPA [1987] showed produced water contain significant levels of BETX, Diesel oil base muds contain BETX but at concentrations low enough to assure attenuation by adsorption onto soil particles [El-Dib et al., 1978]. Hydrocarbon mobility is restricted, further, by chromatographic of liquid moving through a porous media [Warden, Groenewould and Bridie, 1977]. Further, oil floats on water. Accordingly, oil movement through soil is restricted to pores of sufficient diameter, not water saturated. Additionally, the "Jamin effect" [obstruction of a non-wetting fluid flow in porous media] restricts hydrocarbon movement [Schiegg, 1980].

Leaching of hydrocarbons from low concentration contaminated surface soil presents no problem. Watts et al. [1982] observed no oil migration at 30 to 45 cm [12 to 18 inches] depth when 14 percent oily industrial waste was incorporated in the first 15 cm [6 inches] of surface soil. Raymond et al. [1976] found 99 percent of the oil remained within the 20 cm. [8 inches] after 1 year after 2 percent oil was added to the top 15 cm [6 inches] of soil. Streebin et al. [1985] observed no significant oil migration below the zone of incorporation with loading rates of 3 and 13 percent by weight. Oudot [1989] showed only slight tendency of hydrocarbons leaching at a loading of 2 percent in soil. Also, hydrocarbons leaching is reduced in areas of high clay soils. Accordingly, the one-time oil concentrations presented in Table A-I-1 represent defensible limits.

I-14   BIOREMEDIATION

Raymond et al. [1967], Atlas and Bartha [1972], Jobson et al. [1972], Kincannon [1972], Westlake et al. [1974], Horowitz et al. [1975] and Sveinung et al. [1986] observed soils naturally contain adequate diversity of microbes and capacity to biodegrade saturates and light end aromatics [the more mobile fraction]. Concentration and composition of hydrocarbons, nutrients, oxygen, moisture and temperature control the rate of degradation [Schwendinger, 1968; Francke and Clark, 1974; Huddleson and Myers, 1978; Dibble and Bartha, 1979; Brown et al., 1983; Flowers et al., 1984; Bleckmann et al., 1989]. Specifically, the highest rate of bioremediation occurs when:

- carbon/nitrogen ratio = 60–100
- moisture content; 60–80 percent retained by soil at 0.33 bar pressure, and
- temperature = 35–38° C [95–100° F]

Watts et al. [1982] measured a 2 year half life for hydrocarbons from a soil contaminated with 14 percent by weight oil. Streebin et al., [1985] observed a half life of about 2 years at similar loading rate. Oudot et al. [1989] observed 99 percent field biodegradation of the hydrocarbons in soil loaded at 2 percent after 3.5 years. Lynch and Genes [1987] measured a half life of 77 days in field plots contaminated with 5 percent polyaromatic hydrocarbons.

Degradation processes attenuate the mobile, light end aromatics and water soluble petroleum hydrocarbons when applies to the soil surface [Raymond, 1975; Brown et al., 1983; Brown and Deuel, 1983; Whitfell and Boyd, 1987; Bleckmann et al., 1989]. This demonstrates little potential for contaminant migration. Several studies [Plice, 1948; Mackin, 1950; Ellis and Adams, 1961; Baker, 1970;

Giddens, 1976] demonstrate low concentration oil applications improve soil physical conditions and fertility because the degradation product of oil is soil humus.

The limiting threshold criteria depends on land use and method of disposal. These properties are incorporated in the concepts used to develop Table A-I-1.

I-15   RESERVE PIT OPERATION

Drilling fluids function because of chemical additions which deflocculate [disperse] clay particles. This same property results in forming a filter cake on the bore hole surface and the reserve pit surface upon discharge of spent drilling fluids to the pit.

I-15.1   Pit Surface Sealing

Rowsell et al. [1985] observed dispersed particulate material in manure waste formed an effective seal against contaminant migration in unlined, earthen storage impoundments. The mechanism of sealing was a physical blocking of pores by the particulate matter. This same dispersion and physical blockage occurs when drilling fluids are discharged to the reserve pit. Additionally, clay and fine silt incorporated in the mud during formulation or drilling are incorporated within the filter cake formed when excess mud is discharged with the cuttings to the pit. These particles form a natural liner on the pit surface. The "natural liner" forms a physical barrier having chemisorptive properties. Recent unpublished data support the premise that heavy metals, organics and even chlorides do not migrate the "natural liner".

Reserve pits constructed in course textured soils require more time to form a natural liner. Coating the pit surface with bentonite clay provides an initial "seed" surface for forming the natural liner. Additionally, in arid areas containing loamy or clayey soils, a natural liner is slower to form. This allows deeper penetration of pit waste liquids. Prewetting the pit surface after coating with bentonite in sandy soils, or in loamy and clayey soils in arid areas, speeds the formation of a seal, and reduces the depth of penetration and fluid loss.

I-16   SUMMARY OF THRESHOLDS AND APPLICATIONS

Table A-I-4 provides a summary of criteria and applications based on waste type.

E&P pit wastes consist of liquids and solid phases. Solids have utility as construction fill in arid and semi-arid areas. However, pit solids do not make suitable weight bearing or road surfaces without amendments.

Accordingly, roadspreading is not recommended for these solids.

TABLE A-I-4
SUMMARY OF E&P WASTE, DISPOSAL TECHNIQUE, AND OPERATIVE CRITERIA

E&P Waste	Disposal Technique	Criteria			
		EC mmho/cm	SAR	ESP %	O&G %
Liquid	roadspreading	12	NA*	NA*	NA*
	landspreading	4–8	12	15	1
Solids	landspreading	4–8	12	15	1
	burial or landfill	12–24	NA*	NA*	1

NA* - not applicable.

I-17   FLOW DIAGRAMS

I-17.1   Pit Liquids

Pit liquid disposal options depend on the analysis of waste, Figure A-I-1.

I-17.2   Pit Solids

Pit solids evaluation is more complex than required for pit liquids, Figure A-I-2.

I-18   SAMPLE CALCULATIONS

I-18.1   Parameters and Example Calculations for Management of Pit Wastes

The following example demonstrates the calculation method for managing reserve pit wastes by landspreading. The pit contains liquid and solids. Also, sufficient native soil is available for land treatment. Sampling and analysis provides pit and native soil characteristics, Table A-I-5.

I-19   DETERMINATION OF LIMITING CONSTITUENT(S)

I-19.1   Pit Liquid Management

   a) Comparison of pit liquid analyses and threshold values show no chemical limitation for land application.
   b) Native soil loading capacity for Na using an ESP of 12 percent and materials distribution depth of 6 in./acre.

     Given: 1 acre-6 in = 2,000,000 lb.
             1 mg/kg = 1 lb/1,000,000 lb.

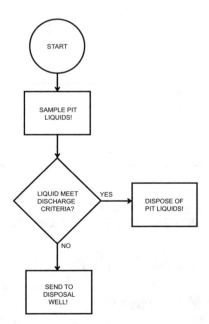

FIG. A-I-1
STEPS FOR PIT LIQUID EVALUATION AND DISPOSAL

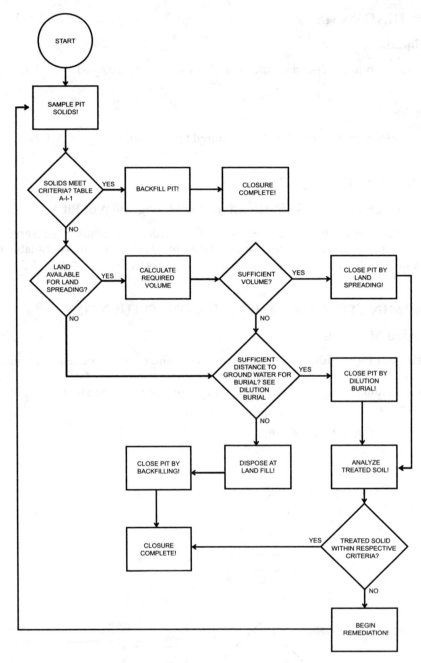

FIG. A-I-2
STEPS FOR PIT SOLIDS EVALUATION AND DISPOSAL

measured CEC = 39.6 meq/100g, Table A-I-5
allowable ESP = 12 percent = 0.12
Na = 23 mg/meq [Annex G]

Then

Na, mg/kg = CEC * ESP * Na
= (39.6 meq/100g) * 0.12 * (23 mg/meq) * 1000

TABLE A-I-5
PIT MATERIAL AND NATIVE SOIL CHARACTERISTICS

Parameter[1]	Pit Liquid[2]	Pit Solids[3]	Native Soil[3]	Threshold Level[4]
Moisture, %	NA*	243.0	NA*	NA*
TDS, mg/liter	1,410	24,830	272	NA*
EC, mmhos/cm	2.3	40.5	0.4	4
SAR, ratio	4	25	<1	12
Na, meq/liter	9.3	260.0	0.4	NA*
Ca, meq/liter	10.9	199.0	2.5	NA*
Mg, meq/liter	1.7	16.0	0.9	NA*
CEC, meq/100g	NA*	13.5	39.6	NA*
Na, meq/100 g	NA*	2.8	0.3	NA*
Ca, meq/100 g	NA*	18.5	24.8	NA*
Mg, meq/100 g	NA*	0.3	7.5	NA*
ESP, %	NA*	20.7	<0.1	12
O&G, %	0.2	10.1	<0.1	1
Volume, bbl	12,938	21,897	NA*	NA*

NA* - not applicable.

[1] Parameters are reported on a dry weight basis unless noted otherwise.

[2] NA - means the parameter meets the guidance threshold or is not applicable for that matrix.

[3] Soluble constituents were determined for saturated paste extracts of pit solids and native soil.

[4] An ESP of 12 percent is recommended in establishing land requirements for Na management.

$$= 39.6 * 0.12 * 23 * 10$$
$$= 1093 \text{ mg/kg}$$

Since 1 mg/kg = 1 lb/1,000,000 lb.

and 1 acre - 6 in = 2.000.000 lbs.

Then

$$\text{Na, lb/acre-6 in} = 1093 * 2 * 10^6/1 * 10^6$$
$$= 2185 \text{ lb/acre-6 in.}$$

c) Mass Na in pit liquids

$$\text{Na} = 9.3 \text{ meq/l}$$

pit volume = 12,938 bbl

$$\text{Na, lb} = \frac{[(9.3 \text{ meq/l}) * (23 \text{ mg/meq}) * (3.8 \text{ l/gal}) * (42 \text{ gal/bbl}) * (12938 \text{ bbl/pit})]}{[(1000 \text{ mg/g}) * (454 \text{ g/lb})]}$$

$$= 973 \text{ lb sodium}$$

d) Land area required sodium distribution, assuming 6 inch depth.

Area = (973 lb Na)/(2186 lb Na/acre-6 in.)

= 0.45 acre - 6 inches

e) Liquid management limitation

12,938 bbl of liquid

$$\text{Area} = \frac{(21,897 \text{ bbl}) * (42 \text{ gal/bbl}) \text{acre ft}^3}{43,560 \text{ ft}^2 * \dfrac{1 \text{ ft}}{12 \text{ in.}} * 7.48 \text{ gal}}$$

= 20 acre-in.

f) Assuming the soil infiltration rate = 1.12 in/hr but drops to less than 0.1 in/hr within 10 minutes. This means that once the surface becomes wet, infiltration is limited by the movement of water or drainage within the profile. The dry surface can receive about 1.2 inches without producing a runoff. The land area required for the pit liquids is:

Area = 20 acre-in/1.2 in
    = 16.7 acres

Constructing a ring levee allows infiltration of liquid at a greater depth over a longer time but using a small area.

I-19.2   Pit Solids Management

a) Compare EC, SAR, ESP and Oil and Grease for pit solids and limiting values. This shows EC, ESP and Oil and Grease are potential limiting constituents. The high calcium content of the solids and receiving soil suggests SAR is not limiting.

b) Percent solids by weight solids analysis shows solids contain 243 percent moisture.

$$\text{Dry weight, g} = \frac{(100 \text{ g wet wt}) * 100}{(100 + 243 \% \text{ moisture})}$$

$$= 29.15 \text{ g}$$

$$\text{Solids, \%} = \frac{29.5 \text{ g}}{100 \text{ g}} * (100) = 29.15\%$$

c) Solids volume

$$\text{Volume, bbl} = 21,897 \text{ bbl wet} * \frac{29.15}{100}$$

$$= 6383 \text{ bbl}$$

d) Total Dissolved Solids
Allowable EC = 4 for land application
        TDS = 4 * 613 = 2452 mg/l

Using proportionately rule, the allowable TDS [2452 mg/l] results from combining native soil and pit solids, on a dry weight basis.

$$2,452 \text{ mg/l} = \frac{(6383 \text{ bbl})(24,830 \text{ mg/l}) + (X \text{ bbl})(272 \text{mg/l})}{(6383 \text{ bbl} + X \text{ bbl})}$$

Solving for X = Volume of native soil

(6383 bbl dry solids) + X bbl native soil) * (2452 mg/l) = (6383 bbl dry solids) * (24,830 mg/l)
                                                    + (X bbl native soil) * (272 mg/l)

Volume of native soil = 65,522 bbl

$$V, \text{acre-6 in} = \frac{65,532 \text{ bbl} * 5.614 \text{ cuft/bbl}}{43,560 \text{ sqft/acre} * \dfrac{\text{ft}}{2 \, (6 \text{ in})}}$$

Volume = 16.89 acre - 6 in.

Thus 16.9 acres are required to reduce the TDS to the acceptable limit.

e) EC Land Requirements
Allowable EC = 4 mmhos/cm for land application.

$$4 \text{ EC} = \frac{(40.5 \text{ EC})(6383 \text{ bbl}) + (0.4 \text{ EC})(X \text{ bbl})}{(6383 \text{ bbl} + X \text{ bbl})}$$

X = 64,717 bbl native soil

$$V, \text{acre - 6 in} = \frac{64,717 \text{ bbl} * 5.614 \text{ cuft/bbl}}{43,561 \text{ sqft/acre} * \dfrac{\text{ft}}{2(6 \text{ in})}} = 16.68 \text{ acre-6 in}$$

Accordingly, 16.7 acre-6 in are required to reduce EC to the acceptable limit.

f) ESP Land Requirement

Allowable ESP = 12 percent

Pit solids ESP = 20.7 percent

$$0.12 * (6383 \text{ bbl dry solids}) + X * 0.12 = (6383 \text{ bbl dry solids}) * (0.207)$$
$$+ (X \text{ bbl native soil}) * (0.001 \text{ mmhos/cm})$$

X = 4667 bbl

   = 1.2 acre-6 in.

Accordingly, only 1.2 acre - 6 in. are required for land treatment of ESP.

g) Oil and Grease Land Requirement

Allowable oil and grease = 1 percent

Pit solids oil and grease = 10.1 percent

Native soil oil and grease = 0.01%

$$1\% \text{ O\&G} = \frac{(10.1\% \text{ O\&G})(6383 \text{ bbl}) + (0.01\% \text{ O\&G})(X \text{ bbl})}{(6383 \text{ bbl} + X \text{ bbl})}$$

X = 64539 bbl

   = 16.7 acre - 6 in.

Accordingly, 16.7 acre - 6 in. are required for land treatment of oil and grease.

h) Limiting Criteria

Comparing the land requirement for EC, ESP and Oil & Grease, it is clear that EC requires the most acreage, 16.7 acre - 6 in. Accordingly, EC is the limiting constituent. Spreading the pit solids over an area consisting of 16.7 acres, allowing the solids to dry and disking and cross-disking to 6 in. satisfies the criteria.

i) Landspreading Solids

Placing wet solids too deep increases the probability of equipment becoming immobilized. Accordingly, spread solids no more than 4 inches thick.

Thickness of

$$\text{wet solids} = \frac{(21,897 \text{ bbl wet solids}(\text{acre} - \text{ft}))}{(16.7 \text{ acre}) * (43560 \text{ cuft})} * \frac{(5.614 \text{ cuft})12 \text{ in.}}{(\text{bbl}) \text{ ft}}$$

   = 2.03 in.

Accordingly, spread wet solids to about 2 inches thick and let dry. Then disk and cross-disk.

j) Nitrogen Addition

The addition of nutrients to the admixed soil encourages plant growth. Add nitrogen in proportion to the carbon content of the admixed soil. As discussed in Chapter 9, oil contains about 78 percent carbon. Phosphorus [P] and Potassium [K] are added at a ratio of N:P:K - 4:1:1. The carbon to nitrogen ratio = 150:1.

$$\text{Oil and grease in the admixture} = 1 \text{ percent oil and grease weight/acre}$$
$$= (1\% \text{ oil and grease}) * (10,000 \text{ ppm/\%}) * 2 \text{ ppm/lb/ac}$$
$$= 20,000 \text{ lb/acre}$$

$$\text{Carbon weight/acre} = (20,000 \text{ lb/acre}) * 0.78$$
$$= 15,600 \text{ lb carbon/acre}$$

N requirement/acre = 15,600/150

$$= 104 \text{ lb/acre}$$

$$\text{For urea (45 percent N) fertilizer required} = \frac{104}{0.45} = 231 \text{ lb/acre}$$

$$\text{For ammonium nitrate (33\%) fertilizer} = \frac{104}{0.33} = 315 \text{ lb/acre}$$

REFERENCES

API, American Petroleum Institute. 1987. Oil and gas industry exploration and production wastes. DOC No. 471-01-09, July.

Atlas, R. M. and R. Bartha, 1972. Degradation and Mineralization of Petroleum by Two Bacteria Isolated From Coastal Waters. Biotechnol. & Bioeng. 14:297–308.

Baker, J. M., 1970. The Effects of Oil on Plants. Environ. Pollution. 1:27–66.

Baldwin, I. L., 1922. Modifications of Soil Flora Induced by Application of Crude Petroleum. Soil Sci. 14:465–477.

Barrow, G. M. 1966. The Nature of Electrolytes in Solution. p. 648–684. Physical Chemistry. Mcgraw-Hill Book Co., New York.

Bleckmann, C. A., L. J. Gawel, D. L. Whitfill, and C. M. Swindoll. 1989. Land Treatment of Oil-Based Drill Cuttings. SPE/IADC Drilling Conf. Proc. New Orleans, LA. p. 529–536.

Bresler, E., B. L. McNeal, and D. L. Carter. 1983. Saline and Sodic Soils. Springer-Verlag Publishing Co. Berlin, Heidelberg, and New York.

Brown, K. W., L. E. Deuel, Jr., 1983. An Evaluation of Subsurface Conditions at Refinery Land Treatment Sites. EPA-600/52-83-096.

Brown, K. W., L. E. Deuel, Jr., and J. C. Thomas, 1983. Land Treatability of Refinery and Petrochemical Sludges. EPA-600/52-83-074.

Bruce, R. R., R. A. Leonard, A. W. Thomas, and W. A. Jackson. 1985. Redistribution of Bromide by Rainfall Infiltration into a Cecil Sandy Loam Landscape. J. Environ. Qual. 14:439–445.

Bulman, T. L. and R. P. Scroggins. 1988. Environment Canada research on land treatment of petroleum wastes. 81st Annual Meeting, APCA. Dallas, Texas.

Carr, R. H., 1919. Vegetable Growth in Soil Containing Crude Petroleum. Soil Sci. 8:67–68.

Deuel, L. E., Jr. and K. W. Brown. 1980. The Feasibility of Utilizing Uranium Leach Water as an Irrigation Source. Final Report. Texas A&M University Research Foundation, College Station, Texas.

Deuel L. E., Jr. and G. H. Holliday. 1993. Determination of Total Petroleum Hydrocarbons in Soil. SPE 26394. October, Houston TX.

Dibble, J. T., and R. Bartha, 1979. Effect of Environmental Parameters on Biodegradation of Oil Sludge in Soil. Soil Sci. 127:365–370.

Duffy, J. J., M. F. Mohtadi, and E. Peake. 1977. Subsurface Persistence of Crude Oil Spilled on Land and Its Transport in Groundwater. p. 475–478. *In*: J. O. Ludwigson (ed.). Proc. 1977 Oil Spill Conf. New Orleans, La. API Washington, D.C.

El-Dib, M. A., A. S. Moursy and M. I. Badawy. 1978. Role of Adsorbents in the Removal of Soluble Aromatic Hydrocarbons from Drinking Waters. Wat. Res. 12:1131–1137.

Ellis, R. E., Jr., and R. S. Adams, 1961. Contamination of Soils by Petroleum Hydrocarbons. Advances in Agron. 13:197–216.

EPA, U. S. Environmental Protection Agency. 1987. Exploration, development, and production of crude oil and natural gas, field sampling and analysis results. EPA 530-SW-87-005, Dec.

Ferrante, J. G., 1981. Fate and Effects of Whole Drilling Fluids and Fluid Components in Terrestrial and Fresh Water Ecosystems. EPA 600/4-81-031.

Flowers, T. H., I. D. Pulford, and H. J. Ducan. 1984. Studies on the Breakdown of Oil in Soil. Environ. Pollut. 8:71–82.

Foth, H. D and L. M. Turk, 1972. Fundamentals of Soil Science, 5th Edition. John Wiley & Sons, Inc., New York, NY.

Frankenberger, W. T., Jr., and C. B. Johanson. 1982. Influence of Crude Oil and Refined Petroleum Products on Soil Dehydrogenase Activity. J. Environ. Qual. 11:602–607.

Francke, H. C. and F. E. Clark, 1974. Disposal of Oily Waste by Microbial Assimilation. U.S. Atomic Energy Report. Y-1934.

Freeman, B. D. and L. E. Deuel, Jr., 1984. Guidelines for Closing Drilling Waste Fluid Pits in Wetland and Upland Areas. 2nd Industrial Pollution Control Proceedings. 7th ETCE Conference, New Orleans, La.

Freeman, B. D. and L. E. Deuel, Jr. 1986. Closure of freshwater base drilling mud pits in wetland and upland areas. Proceedings of a National Conference on Drilling Muds. Univ. of Okla. Environ. and Groundwater Institute. Norman, Okla. May 29–30. p. 303–327.

Garner, J. H., 1971. Changes in Soil and Death of Woody Ornamentals Associated with Heating Natural Gas. Phytopathology 61:892.

Giddens, J., 1976. Spent Motor Oil Effects on Soil and Crops. J. Environ. Qual. 5:179–181.

Harper, J. T., 1939. The Effect of Natural Gas on the Growth of Microorganisms and Accumulation of Nitrogen and Organic Matter in the Soil. Soil Sci. 48:461–466.

Haywood, H. E. and C. H. Wadleigh. 1949. Plant growth on saline and alkali soils. Advan. in Agron. 1:1–38. Agron. Soc. of Amer. Madison, Wis.

Hem, J. D. 1985. Study and interpretation of chemical characteristics of natural water. 3rd Edition. U.S. Geological Survey Water Supply Paper 2254. Alexandria, VA.

Henderson, G. 1982. Analysis of hydrologic and environmental effects of drilling mud pits and produced water impoundments. Dames and Moore, Houston, Texas. Report available from API, Dallas, Texas.

Horowitz, A., D. Gutrick, and E. Rosenberg, 1975. Sequential Growth of Bacteria on Crude Oil. Appl. Microbiol. 30:10–19.

Huddleston, R. L. and J. D. Myers, 1978. Treatment of Refinery Oily Wastes by Land Farming. 85th Natl. Meeting of Amer. Inst. Chem. Eng. Philadelphia, Pa. 4–8 June, 1978. Amer. Inst. Chem. Eng., New York, NY.

Jobson, A., F. D. Cook, and W. S. Westlake, 1972. Microbial Utilization of Crude Oil. Appl. Microbiol. 24:1082–1089.

Kincannon, C. B. 1972. Oily Waste Disposal by Cultivation Process. Office of Research and Monitoring, Rep. No. EPA-R2-72-110. Washington, D.C., 20460.

Kramer, P. J. 1969. Plant and Soil Water Relationships: A Modern Synthesis. McGraw-Hill Inc., New York.

Lunin, J. 1967. Water for Supplemental Irrigation. p. 63–78. *In*. Water Quality Criteria, ASTM STP 416. 1st National Meeting on Water Quality Criteria, Philadelphia, Pa. September.

Lynch, J. and B. R. Genes. 1987. Land treatment of hydrocarbon contaminated soils. p. 163–174. *In*. P. T. Kostecki and E. J. Calabrese (eds.) Petroleum contaminated soils. Volume I. Remediation techniques environmental fate risk assessment. Lewis Publishers, Chelsea, MI.

Maas, E. V. and G. J. Hoffman, 1977. Crop Salt Tolerance - Current Assessment. Journal of the Irrigation and Drainage Division Proceedings, Amer. Soc. of Civil Eng. Vol 103 no. IRZ.

Maas, E. V. 1986. Salt tolerance of plants. Appl. Agric. Res. 1:12–26.

Mackin, J. G. 1950. Report on a Study of the Effects of Applications of Crude Petroleum on Salt Grass, Distichlis spicota. L. Green, Texas A&M Res. Foundation Project 9, p 8–54.

Miller, R. W., and S. Honarvar, 1975. Effect of Drilling Fluid Component Mixtures on Plants and Soil. *In* Environmental Aspects of Chemical Use in Well Drilling Operation, Conference Proceedings, May, Houston, TX. EPA 560/1-75-004 p. 125–143.

Miller, R. W., S. Honarvar, and B. Hunsaker. 1980. Effects of Drilling Fluids on Soils and Plants: I. Individual Fluid Components. J. Environ. Qual. 9:547–552.

Miller, R. W., and P. Pesaran. 1980. Effects of Drilling Fluids on Soil and Plants: II. Complete Drilling Fluid Mixtures. J. Environ. Qual. 9:552–556.

Murphy, H. F. 1929. Some effects of Crude Petroleum on Nitrate Production, Seed Germination and Growth. Soil Sci. 27:117–120.

Murphy, E. C. and A. E. Kehew. 1984. The effect of oil and gas well drilling fluids on shallow groundwater in western North Dakota. Report No. 82. North Dakota Geological Survey.

Nelson, M., S. Liu, and L. Sommers, 1984. Extractability and Plant Uptake of Trace Elements from Drilling Fluids. J. Environ. Qual. 13:562–566.

Odu, C. T. I., 1972. Microbiology of Soils Contaminated with Petroleum Hydrocarbons. Extent of Contamination and Some Soil and Microbiological Properties After Contamination. J. Inst. Petrol. 58:210–208.

Overcash, M. R. and D. Pal. 1979. Design of land treatment systems for industrial waste - theory and practice. Ann Arbor Sci., Ann Arbor, Mi.

Oster, J. D. and J. D. Rhoades. 1984. Water management for salinity and sodicity control. Chapter 7, p. 1–20. *In*. Irrigation with reclaimed municipal wastewater. Lewis Publishers, Chelsea, MI.

Oudot, J., A. Ambles, S. Bourgeois, C. Gatellier, and N. Sebyera. 1989. Hydrocarbon infiltration and biodegradation in land farming experiment. Environ. Pollut. 59:17–40.

Owens, L. B., R. W. Van Kuren, and M. M. Edwards. 1985. Groundwater Quality Changes Resulting from a Surface Bromide Application to a Pasture. J. Environ. Qual. 14:543–548.

Pal, D., and M. R. Overcash, 1978. Plant-Oil Assimilation Capacity for Oils. 85th Natl. Meeting of Amer. Inst. Chem. Eng. Philadelphia, Pa.

Patrick, Z. A. 1971. Phytotoxic substances associated with the decomposition in soil of plant residues. Soil Sci. 111:13–18.

Plice, M. J., 1948. Some Effects of Crude Petroleum on Soil Fertility. Soil Sci. Soc. Amer. Proc. 13:413–416.

Raymond, R. L., V. W. Jamison and J. O. Hudson, 1967. Microbial Hydrocarbon Co-oxidation of Mono and Dicyclic Hydrocarbons by Soil Isolates of the Genus *Norcardia*. Appl. Microbiol. 15: 357–865.

Raymond, R. L., J. O. Hudson, and V. W. Jamison, 1975. Assimilation of Oil by Soil Bacteria. *In* Proc. 40th Annual Midyear Meeting API Refining.

Raymond, R. L., J. O. Hudson, and V. W. Jamison. 1976. Oil degradation in soil. Applied and Environ. Microbiol. 31:522–535.

Reeve, R. C. and M. Fireman. 1967. Salt problems in relation to irrigation. *In.* R.C. Dinauer (ed.) Irrigation of agricultural lands. Agron. 11:908–1008. Am. Soc. of Agron., Madison, Wis.

Rowsell, J. G., M. H. Miller, and P. H. Groenevelt. 1985. Self-Sealing of Earthen Liquid Manure Storage Ponds: II. Rate and Mechanism of Sealing. J. Environ. Qual. 14:539–543.

Roy, W. R. and R. A. Griffin. 1985. Mobility of Organic Solvents in Water-Saturated Soil Materials. Environ. Geol. Wat. Sci. 7:241–247.

Schiegg, H. O. 1980. Field Infiltration as a Method for the Disposal of Oil-in-Water Emulsions from the Restoration of Oil-Polluted Aquifers. Wat. Res. 14:1011–1016.

Schollenberger, C. J., 1930. Effect of Leaking Natural Gas upon the Soil. Soil Sci. 29:260–266.

Schwendinger, R. B., 1968. Reclamation of Soil Contaminated with Oil. J. Inst. Petrol. 54:182–192.

Streebin, L. E., J. M. Robertson, H. M. Schornick, P. T. Bowen, K. M. Bagawandos, A. Habibafsan, T. G. Sprehe, A. L. Callender, C. J. Carpenter, and V. G. McFarland. 1985. Land treatment of petroleum refinery sludges. EPA 600/2-84-193.

Sveinung, S., A. Lode, T. A. Pedersen. 1986. Biodegradation of Oily Sludge in Norwegian Soils. Appl. Microbiol. Biotechnol. 23:297–301.

Thoresen, K. M. and A. A. Hinds. 1983. A Review of the Environmental Acceptability and the Toxicity of Diesel Oil Substitutes in Drilling Fluid Systems. IADC/SPE Drilling Conf. Proc. New Orleans, LA. p. 343–352.

Tucker, B. B. 1985. Soil application of drilling wastes. Proceedings of National Conference on Disposal of Drilling Wastes. Univ. of Okla. Environ. and Groundwater Institute. Norman, Okla. May 29–30. p. 102–112.

Udo, E. J. and A. A. A. Fayemi. 1975. The effect of oil pollution of soil on germination, growth and nutrient uptake of corn. J. Environ. Qual. 6:369–372.

U. S. Salinity Laboratory Staff, 1954. Diagnosis and Improvement of Saline and Alkali Soils. Agric. Handb no. 60, USDA. U.S. Government Printing Office, Washington, DC.

Waarden, M. Van Der, W. M. Groenewoud, A. L. A. M. Bridie. 1977. Transport of Mineral Oil Components to Groundwater II. Wat. Res. II. 359–365.

Watts, J. R., J. C. Corey, and K. W. McLeod. 1982. Land application studies of industrial waste oils. Environ. Pollut.(Series A) 28:165–175.

Weldon, R. A. 1978. Biodegradation of Oily Sludge by Soil Microorganisms. Suntech Project 48-036. Rep. No. 2, API Washington, D.C.

Westlake, D. W. S., A. Jobson, R. Phillippe, and F. D. Cook, 1974. Biodegradability and Crude Oil Composition. Can. J. Microbiol. 20:915–928.

Whitfill, D. L. and P. A. Boyd. 1987. Soil Farming of Oil Mud Drill Cuttings. SPE/IADC Drilling Conf. Proc. New Orleans, LA. p. 429–438.

Younkin, W. E. and D. L. Johnson. 1980. The impact of waste drilling fluids on soils and vegetation in Alberta. Symp. Proc. Vol. 1: Research on Environmental Fate and Effects of Drilling Fluids and Cuttings. Lake Buene Vista, Fl. Jan. 21–24. p. 98–112.

U. S. EPA 40 CFR Part 143 - National Secondary Drinking Water Regulations. Section 143.3 - Secondary maximum contaminant levels. July 1, 1984.

ANNEXES

ANNEX A : OIL & GREASE

ANNEX B : EC

ANNEX C : TDS

ANNEX D : CATION EXCHANGE CAPACITY

ANNEX E : E&P SAMPLE PREPARATION

ANNEX F : EXCHANGEABLE CATIONS

ANNEX G : SAR

ANNEX H : SATURATED PASTE EXTRACT

ANNEX A
OIL & GREASE

1.0. Scope and Applications
   1.1. Use this method to recover O&G by chemically drying wet E&P waste solids and then extracting by Soxhlet apparatus.

2.0. Summary of Method
   2.1. Anhydrous sodium sulfate is used to combine with water and enhance recovery of petroleum hydrocarbon. After drying, the O&G is extracted with methylene chloride using the Soxhlet apparatus.

3.0. Apparatus and Materials
   3.1. Soxhlet extraction apparatus
   3.2. Analytical balance
   3.3. Extraction thimble
   3.4. Grease free glass wool
   3.5. Vacuum distilling apparatus
   3.6. Desiccator

4.0. Reagents
   4.1. Concentrated hydrochloric acid
   4.2. Anhydrous sodium sulfate
   4.3. Nanograde methylene chloride

5.0. Procedure
   5.1. Weigh 25 g ($\pm$ 0.5g) of wet E&P waste solid of soil into 150 ml beaker.
   5.2. Acidify to pH 2 with concentrated hydrochloric acid.
   5.3. Add anhydrous sodium acetate as necessary to dry solids.
   5.4. Transfer sample to extraction thimble, covering sample with glass wool, then place in Soxhlet apparatus.
   5.5. Add methylene chloride and commence extraction at 20 cycles/hr for a minimum of 6 hr.
   5.6. Using grease free glass wool filter extract into a pre-weighed boiling flask, previously rinsed with solvent.
   5.7. Connect boiling flask to vacuum distillation head and evaporate solvent.
   5.8. Place boiling flask in a dessicator to cool and remove traces water on glass.
   5.9. Weigh boiling flask and record weight gain.

6.0. Calculations
   6.1. O&G
   O&G, % = (weight gain in flask, g) / (sample wt, g) $\times$ 100
   where: sample wt. g = (wet weight $*$ 100) / (100 + % moisture)

7.0. References
   7.1. Test Methods for Evaluating Solid Waste. 1986. Method 3540. Soxhlet Extraction. EPA SW-846. USEPA Washington D.C.

ANNEX B
EC

1.0. Scope and Application
   1.1. Electrical conductivity indicates the quantity of soluble salts in an aqueous sample. This method applies to pit liquids and saturated paste extracts.

2.0. Summary of Method

2.1. EC is measured directly with the reading corrected to specific conductance at 25°C.

3.0. Apparatus and Materials

3.1. Temperature compensating conductivity meter

3.2. Conductivity cell

3.3. Reagents

3.3.1. ASTM Type II water

3.3.2. 0.01 N potassium chloride

4.0. Procedure

4.1. Rinse conductivity cell and fill with calibration standard. Read and record conductivity.

4.2. Rinse conductivity cell and fill with sample. Read and record conductivity.

5.0. Calculations

5.1. Cell Constant, C

$$C = (1.413 \text{ mmhos/cm}) / (EC_{KCl} \text{ mmhos/cm})$$

where

EC_{KCl} = measured conductance, mmhos/cm

5.2. Specific Conductance of Sample

$$EC = (EC_m)(C)$$

where

measured conductance of sample, mmhos/cm

C = cell constant

6.0. References

6.1. Rhoades, J.D. 1982. Soluble Salts. p. 172–173. *In* A.L Page (ed.) Methods of Soil Analysis. Part 2 - Chemical and Microbiological Properties. 2nd Edition. (Ed.) ASA Agronomy Monograph 9.

ANNEX C
TDS

1.0. Scope and Application

1.1. This procedure applies to E&P aqueous phase samples including produced water, pit liquids and saturated paste extracts.

2.0. Summary of Method

2.1. Total Dissolved Solid is mineral matter passing a standard glass filter, and remaining after drying at 180°C to constant weight.

3.0. Interferences

3.1. The principle interference is from fine clay fractions and organic colloids passing the filter and stable at 180°C.

4.0. Apparatus and Materials

4.1. Evaporating dishes

4.2. Filtration equipment

4.3. 0.45-μm filters

4.4. Drying oven, for operation to 180°C (+/- 2°C)

4.5. Analytical balance, capable to 0.1 mg

5.0. Procedure
    5.1. Assemble filtration equipment and insert 0.45 μm- filter.
    5.2. Apply vacuum and wash disk with 3, 20-ml volumes of distilled water. Discard washwater.
    5.3. Filter measured volume of homogenized sample through filter, wash with 3, 10-ml volumes of distilled water, allowing complete drainage between washings.
    5.4. Transfer filtrate to weighed evaporation dish previously cleaned by ignition to 550°C for 1 hr.
    5.5. Evaporate water at 180°C to a constant weight. Evaporation dish is cooled in desiccator prior to weighing.

6.0. Calculation

$$\text{TDS, mg/liter} = (A - B) \times 1000/\text{sample volume, ml}$$

where
    A = weight of residue + dish, mg
    B = weight of dish, mg

7.0. References
    7.1. Standard Methods for the Examination of Water and Wastewater. 1985. 16th Edition. APHA. AWWA. WPCF. Method 209 B. Total Dissolved Solids Dried at 180C.

ANNEX D
CATION EXCHANGE CAPACITY

1.0. Scope and Application
    1.1. This method applies to most soils and E&P waste, including calcareous and non-calcareous samples.

2.0. Summary of Method
    2.1. The sample is saturated with an excess of sodium acetate solution, resulting in an exchange of other cations by sodium. Subsequently, excess sodium is rinsed from the sample followed by quantitative desorption of sodium by ammonium. The concentration of displaced sodium is then determined by atomic absorption, emission spectroscopy, or an equivalent means as available and approved by EPA.

3.0. Interferences
    3.1. Soluble salts and gypsum will interfere with the CEC determination if they are present in sufficient quantities. These may be overcome by washing the solids with water before saturating with sodium, or employ a more exhaustive saturation procedure.

4.0. Apparatus and Materials
    4.1. Centrifuge and centrifuge tubes
    4.2. Mechanical shaker
    4.3. Volumetric flask: 100 ml
    4.4. Atomic absorption or equivalent instrumentation

5.0. Reagents
    5.1. Sodium acetate 1.0 N buffered to pH 8.2
    5.2. Ammonium acetate 1.0 N buffered to pH 7.0
    5.3. Isopropyl alcohol: 99%
    5.4. Sodium standards in 1.0 N ammonium acetate

6.0. Sample Preparation
    6.1. See E&P Sample Preparation

7.0. Procedure
    7.1. Weigh 5 g sample into a 50-ml centrifuge tube.

7.2. Add 30 ml of 1.0 N sodium acetate, stopper and shake for 5 min., then centrifuge to clear supernatant.

7.3. Decant and discard supernatant, and repeat step 7.2 three more times to effect sodium saturation.

7.4. Add 30 ml of 99% isopropyl alcohol, stopper and shake for 5 min., then centrifuge to clear supernatant.

7.5. Decant alcohol and discard supernatant, and repeat step 7.4 three more times to effect washing of solids.

7.6. Add 30 ml of ammonium acetate, stopper and shake 5 min., then centrifuge to clear supernatant liquid. Decant supernatant into a 100-ml volumetric flask.

7.7. Repeat step 7.6 two more times decanting into the same volumetric flask.

7.8. Dilute the volumetric to mark with ammonium acetate, and determine sodium concentration by atomic absorption or other instrumentation

8.0. Calculations

8.1. CEC

$$\text{CEC, meq/100 g} = (\text{sodium, meq/litter} \times 10) / (\text{sample wt., g})$$

8.2. ESP

$$\text{ESP, \%} = \frac{(\text{Exchangeable Sodium, meq/100 g})}{(\text{CEC, meq/100 g})} * 100$$

9.0. References

9.1. Chapman, H.D. 1965. Cation Exchange Capacity. p. 891–900. *In* C. A. Black (ed.) Methods of Soil Analysis. Part 2-Chemical and Microbiological Properties. ASA Agron. Monograph 9.

ANNEX E
E&P SAMPLE PREPARATION

1.0. Scope and Application

1.1. Use this method to prepare samples for analysis by the protocols listed below:

1.1.1. Sodium Adsorption Ratio [SAR]

1.1.2. Exchangeable Sodium Percentage [ESP]

1.1.3. Cation Exchange Capacity [CEC]

2.0. Summary of Method

2.1. Homogenize the sample, dried at 105°C and ground prior to the individual analyses.

3.0. Apparatus and Materials

3.1. Oven capable to 105°C (± 2°C)

3.2. Grinding apparatus

3.3. Drying pans

3.4. Balance

4.0. Procedure

4.1. Homogenize the sample thoroughly.

4.2. Weigh to the nearest 0.1 g a pan large enough to hold 250 g sample.

4.3. Weigh 100 to 200 g homogenized sample to pan, and place pan in oven at 105°C until a constant weight is achieved. Record weights to calculate moisture content.

4.4. Grind the material so it will pass a 2-mm sieve. Sample is now ready for analyses.

5.0. Procedure for Hydrophobic Material
 5.1. Tests for hydrophobicity
  5.1.1. Visible blobs of oil or grease
  5.1.2. Sample presses into a single damp looking mass when crushed with mortar and pestle and will not hydrate with water.
  5.1.3. Sample leaves an oily mark when pressed between two pieces of filter paper.
  5.1.4. Sample feels damp when pinched between fingers.
 5.2. Place sample in muffle furnace and heat to 250°C for 1 hr.
 5.3. Raise temperature to 350°C at 50°C intervals allowing smoke to dissipate between adjustments. Do not allow sample to catch fire or exceed 390°C.
 5.4. Cool the sample and grind it to pass 2-mm sieve. The sample is now ready for the appropriate analyses.

6.0. Calculation
 6.1. Moisture Content (dry weight basis)
  Based on weights from steps 4 or 5

$$\text{Moisture, } \% = (W - D)/(D - P) * 100$$

where

W = wet weight of sample + pan, g
D = dry weight of sample + pan, g
P = weight of pan, g

ANNEX F
EXCHANGEABLE CATIONS

1.0. Scope and Application
 1.1. This method applies to most soils and E&P waste solids and determines the distribution of cations adsorbed on the solid phase.

2.0. Summary of Method
 2.1. The sample is saturated with an excess of ammonium acetate resulting in an exchange of adsorbed cations. The cations released into solution are then quantified as extractable cations and when adjusted for soluble cations are reported as exchangeable cations.

3.0. Interferences
 3.1. Sparingly soluble salts may give erroneously high cation distribution values.

4.0. Apparatus and Materials
 4.1. Centrifuge and centrifuge tubes
 4.2. Mechanical shaker
 4.3. Atomic absorption or other suitable instrumentation

5.0. Procedure
 5.1. Weigh 5 g of sample to a 50 ml centrifuge tube.
 5.2. Add 30 ml 1N ammonium acetate reagent to the tube, stopper, and shake for 5 min. and centrifuge to yield a clear supernatant liquid.
 5.3. Decant the supernatant as completely as possible into a 100 ml volumetric flask. Repeat step
 5.2. two more times combining extracts.
 5.4. Dilute to volume [100 ml], mix, and determine the amounts of the various extracted cations using AAS or other suitable instrumentation.
 5.5. Soluble cations must be determined for an aqueous extract of the same sample if not determined previously.

6.0. Calculations

   6.1. Extractable Cations

$$\text{extractable cation, meq/100 g} = \frac{(\text{cation concentrat ion of extract in meq/liter}) * 10}{(\text{sample wt in g.})}$$

   6.2. Soluble Cations

$$\text{soluble cation, meq/100 g} = (\text{cation concentration of saturation extract in meq/liter}) * (\text{saturation percentage}) / 1000$$

   6.3. Exchangeable Cations

$$\text{exchangeable cation, meq/100 g} = (\text{extractable cation in meq/100 g}) - (\text{soluble cation in meq/100 g})$$

7.0. References

   7.1. Thomas, G.W. 1982. Exchangeable Cations. p. 159–161. *In* A.L. Page (ed.) Methods of Soil Analysis. Part 2 - Chemical and Microbiological Properties. 2nd. Edition. ASA Agron. Monograph 9.

ANNEX G
SAR

1.0. Scope and Application

   1.1. This method applies to most E&P wastes including pit liquids and water extracts of pit solids or waste solid:soil mixtures.

2.0. Summary of Method

   2.1. Soluble cations are determined by atomic absorption spectrophotometry or other suitable instrumentation for pit liquids or water extracts of solid phase samples. The sodium adsorption ratio (SAR) is calculated from the cationic distributions.

3.0. Procedure

   3.1. Calibrate instrumentation using standards of known concentration.

   3.2. Read concentrations of Na, K, Mg and Ca direct for pit liquid samples or aqueous extracts including saturated pastes.

4.0. Calculations

   4.1. Conversion to meq/liter

      Na, meq/liter = (Na mg/liter) / (23 mg/meq)

      K, meq/liter = (K mg/liter) / (39 mg/meq)

      Ca, meq/liter = (Ca mg/liter) / (20 mg/meq)

      Mg, meq/liter = (Mg mg/liter) / (12 mg/meq)

   4.2. SAR

$$\text{SAR} = (\text{Na, meq/l}) / [(\text{Ca, meq/l} + \text{Mg, meq/l})/2]^{1/2}$$

5.1. References

   5.1. Rhoades, J.D. 1982. Soluble Salts. p. 173–174 A.L Page (ed.) Methods of Soil Analysis. Part 2 - Chemical and Microbiological Properties. 2nd. Edition. ASA Agro. Monograph 9.

ANNEX H
SATURATED PASTE EXTRACT

1.0 Scope and Applications

   1.1. Saturation percentage is a condition of soil related to field moisture and associated plant response. It is reproducible and approximately equivalent to twice the percent moisture at field capacity (0.33 bar) and 4 times the percent moisture at permanent wilting (15 bar). This method is used to obtain a saturation extract for the following analyses:

1.1.1. TDS

1.1.2. EC

1.1.3. SAR

2.0. Summary of Method

2.1. Add water to a known amount of sample until point where no more water can be added without forming free water layer.

3.0. Interferences

3.1. Excessive stirring puddles the sample and reconstitute the dispersed condition of most E&P waste solids. Puddled soils represent a gross over estimation of the saturation percentage.

4.0. Apparatus

4.1. Container of 250 ml capacity.

4.2. Buchner funnel, filter paper, vacuum source, and collection vessel.

5.0. Procedure

5.1. Weigh 100 g, dried, ground and sieved solids into 250 ml container.

5.2. Add distilled water to fill pores, stirring gently as needed to achieve saturation. The solid: water mixture is consolidated occasionally by tapping container on workbench.

5.3. At saturation the mixture glistens as it reflects light, flows slightly when the container is tipped.

5.4. Allow paste mixture to stand 1 hr. and check for conditions of paste. Mixture should not stiffen nor should free water form at the surface.

5.5. Add a known quantity of solid sample material if free water forms or more distilled water if mixture stiffens.

5.6. Record the weight of water used to achieve saturation and transfer to the vacuum filter apparatus. Vacuum extraction should be terminated when air begins to pass through the filter.

5.7. Extract is used to measure TDS, EC and SAR.

6.0. Calculation

Saturation Percentage (SP) % = $(W - D)/(D - P) * 100$

where

W = wet weight of sample + pan, g

D = dry weight of sample + pan, g

P = weight of pan, g

7.0. References

7.1. U. S. Salinity Laboratory Staff. 1954. Diagnosis and improvement of saline and alkali soils. Agriculture Handbook 60.

APPENDIX II
E&P TOTAL METALS ANALYSIS

II-1   INTRODUCTION

The analytical methods presented below follow the Louisiana 29-B protocol, but we modified the procedures to apply to soils and wastes in any locale.

We strongly recommend use of these procedures since the limiting criteria herein contained were derived using the prescribed protocols. Additionally, many of the analytical methods presented here have not been adopted as concise standards. Accordingly, no other documented protocols are available.

II-2   ACID DIGESTION OF SLUDGES

II-2.1   Scope and Application

This method is an acid digestion procedure used to prepare sediments, sludges and soil samples for analysis by flame or furnace atomic adsorption spectroscopy (FLAA and GFAA, respectively) or by inductively coupled argon plasma spectroscopy (ICP).

Samples prepared by this method may be analyzed by ICP for all the listed metals, or by FLAA or GFAA as indicated below (See Summary of Method). Mercury is analyzed by cold vapor technique.

FLAA	GFAA	HYDRIDE/COLD VAPOR
Barium	Arsenic	Arsenic
Cadmium	Cadmium	Selenium
Chromium	Chromium	Mercury
Lead	Selenium	
Silver		
Zinc		

II-2.2   Summary of Method

A representative 1g sample is digested in nitric acid and hydrogen peroxide. The digestate is then refluxed with nitric acid or hydrochloric acid. Dilute hydrochloric acid is used as the reflux acid for (1) the hydride analysis of As and Se, and (2) the flame AA or ICP analysis of Ba, Ca, Cd, Cr, Pb, and Zn.

155

II-2.3   Interferences

Sludge samples can contain diverse matrix types, each of which may present its own analytical challenge. Spiked samples and any relevant standard reference material should be processed to aid in determining whether this method is applicable to a given waste.

Hydride Interferences:

a) High concentrations of chromium, cobalt, copper, mercury, molybdenum, nickel, and silver can cause analytical interferences.

b) Traces of nitric acid left following the work-up can result in analytical interferences. Nitric acid must be distilled off by heating the sample until fumes of SO_3 are observed.

c) Elemental arsenic and selenium and many of their compounds are volatile; therefore, certain samples may be subject to losses during sample preparation.

II-2.4   Apparatus and Materials

Conical Phillips beakers: 250-ml
Watch glasses.
Thermometer: That covers range of 0 to 200 C.
Whatman No. 41 filter paper (or equivalent).
Centrifuge and centrifuge tubes.

II-2.5   Reagents

ASTM Type II water (ASTM D-1193): Water should be monitored for impurities.

Concentrated nitric acid, reagent grade (HNO_3): Acid should be analyzed to determine level of impurities. If method blank is <MDL, the acid can be used.

Hydrogen peroxide (30% H_2O_2: Oxidant should be analyzed to determine level of impurities.

II-2.6   Sample Collection, Preservation and Handling

All samples must have been collected using a sampling plan that results in a representative sample.

All sample containers must be prewashed with detergents, acids, and type II water. Plastic and glass containers are both suitable.

II-2.7   Procedure

1.1 Mix the sample thoroughly to achieve homogeneity. For each digestion procedure, weigh to the nearest 0.1 mg and transfer to a conical beaker a 1g portion of sample.

1.2 Add 10-ml of 1:1 HNO_3, mix the slurry, and cover with a watch glass. Heat the sample to 95 C and reflux for 10 to 15 min without boiling. Allow the sample to cool, add 5-ml of concentrated HNO_3, replace the watch glass, and reflux for 30 min. Repeat this last step to ensure complete oxidation. Using a ribbed watch glass, allow the solution to evaporate to 5-ml without boiling, while maintaining a cover of solution over the bottom of the beaker.

1.3 After step 1.2 has been completed and the sample has cooled, add 2-ml type II water and 3-ml of 30% H_2O_2. Cover the beaker with a watch glass and return the covered beaker to the hot plate for warming and to start the peroxide reaction. Care must be taken to ensure the losses do not occur due to excessively vigorous effervescence. Heat until effervescence subsides and cool the beaker.

1.4 Continue to add 30% H2O2 in 1-ml aliquots with warming until the effervescence is minimal or until the general sample appearance is unchanged.

Note: Do not add more than 10-ml 30% H_2O_2 total.

1.5 If the sample is being prepared for (a) the hydride analysis of As and Se, or (b) the flame AA or ICP analysis of Ba, Ca, Cd, Cu, Pb and Zn, then add 5-ml of concentrated HCL and 10 ml of Type II water, return the covered beaker to the hot plate, and reflux for an additional 15 min without boiling. After cooling, dilute 100-ml with Type II water. Particulates in the digestate that may clog the nebulizer should be removed by filtration, by centrifugation or by allowing the sample to settle.

  a) Filtration: Filter through Whatman No. 41 filter paper (or equivalent) and dilute to 100-ml with Type II water.

  b) Centrifugation: Centrifugation at 2,000-3,000 rpm for 10 min is usually sufficient to clear the supernatant of particulates.

  c) The diluted sample has an approximate acid concentration of 5.0% (v/v) HCL and 5.0% (v/v) HNO_3. The sample is now ready for analysis.

1.6 If the sample is being prepared for the furnace analysis of As, Cd, Cr, Pb, and Se cover the sample with a ribbed watch glass and continue heating the acid-peroxide digestate until the volume has been reduced to approximately 5-ml. After cooling, dilute to 100-ml with Type II water. Particulates in the digestate should be removed by filtration, centrifugation, or by allowing the particles to settle.

  a) Filtration: Filter through Whatman No. 41 filter paper (or equivalent) and dilute to 100-ml with Type II water.

  b) Centrifugation: Centrifuge at 2,000–3,000 rpm for 10 min to clear supernatant.

  c) The diluted digestate solution contains approximately 5% (v/v) HNO_3. For analysis, withdraw aliquots of appropriate volume and add any required reagent or matrix modifier.

II-2.8   Calculations

The concentrations determined are to be reported on a dry weight basis. A correction is made for the moisture content.

Moisture content is determined on a separate subsample by heating the sample to 105 C.

$$\text{Moisture, \%} = [(\text{wet wt} - \text{dry wt})/\text{dry wt.}] * 100$$

II-2.9   Quality Control

For each group of samples processed, preparation blanks (Type II water and reagents) should be carried through the entire sample preparation and analytical processes. These blanks will be useful in determining if samples are being contaminated.

Duplicate samples should be processed on a routine basis. Duplicate samples will be used to determine precision. The sample load will dictate the frequency, but 20% is recommended.

Spiked samples and standard reference materials are used to determine accuracy in analysis. A spiked sample should be included with each group of samples processed and whenever a new sample matrix is being analyzed.

The concentration of all calibration standards should be verified against a quality control check sample obtained from an outside source.

II-2.10   References

U.S. Environmental Protection Agency. SW-846 2nd Ed., September, 1990.
Test Methods for Evaluating Solid Waste. Office of Solid Waste and Emergency Response. USEPA, Washington D.C.

II-3   TOTAL BARIUM

II-3.1   Scope and Application

This method is used to determine the total amount of barium in soils and solid wastes containing barite.

II-3.2   Summary of Method

The sample is dried and pulverized to pass a 100 mesh sieve and digested in nitric acid. Atomic absorption or ICP emission spectroscopy is used to determine the barium concentration.

II-3.3   Interferences

Since barite (barium sulfate) is only sparingly soluble in acid, the analyst must strictly adhere to the size of the sample and the volume of acid that is specified in the protocol.

II-3.4   Apparatus and Materials

Erlenmeyer flask, 500-ml
Mortar and pestle, agate
Nitric acid (3+7)
100 mesh sieve
Volumetric flask, 500-ml

II-3.5   Procedure

1.1 Prepare the sample as in the sample preparation section.
1.2 Grind a 10g sample to pass a 100 mesh sieve.
1.3 Weigh 100 mg ± 0.1 mg sample into 500-ml flask and add 400-ml Type II water.
1.4 Reflux on a hot plate until the volume in 300-ml
1.5 Transfer to a 500-ml volumetric flask and bring up to the mark.
1.6 Determine the barium concentration by AA or ICP emission spectroscopy.
1.7 Report the barium content as parts per million on a dry weight basis

APPENDIX III
FIELD SAMPLING FOR LOCATING GROUNDWATER TABLES

III-1 INTRODUCTION

Depth to groundwater is a concern for NOW management techniques utilizing burial and vertical dilution. The concern emanates from the potential for materials buried or distributed in a subsurface profile to be brought back to the surface or further concentrated by a rising water table. For this reason it is suggested that the depth of the trench or burial vault be maintained a minimum 5 ft above the seasonal high groundwater table. The state of Louisiana restricts this activity by statute (Amendment to Statewide Order No. 29-B).

Perhaps more obvious is the potential for direct dissolution of waste constituents into groundwater on contact. The concern is possible contamination of shallow aquifers used for human and animal consumption.

Weathered soils in humid regions may have a saturated zone develop at a discontinuous lithic contact during periods of high rainfall. The discontinuity could be either textural (i.e. sand over clay) or structural (i.e. weakly aggregated over massive) in nature. The saturated zone developed is called a perched water table. A perched water table is separated from permanent groundwater by an unsaturated or "vadose" layer. A sufficiently thick unsaturated zone presents no limitation for NOW management via burial techniques.

This paper advances a procedure for determining moisture regimes with depth. The procedure utilizes incremental physical and moisture analyses on continuous cores to describe a given site in vertical profile. Parameters analyzed include 15 bar moisture, 1/3 bar moisture, the existing moisture content, and porosity.

The 15 bar moisture content describes a condition in which water is bound to soil particles so tight that it is unavailable to plants. Hydraulic gradients and water transport mechanisms are controlled by diffusivity and other unsaturated flow phenomenon. It is measured by first saturating the soil with water, then allowing it to come to equilibrium with an applied pressure of 15 bars.

The 1/3 bar moisture content represents the maximum volume of water a soil can hold before free liquid subject to gravity flow forms. Soil is saturated, then allowed to come into equilibrium with an applied pressure of 1/3 bars.

Moisture as received is a gravimetric determination of water content existing at the time the sample was collected. Porosity is simply the the percentage of open spaces or voids in a soil and represents the maximum moisture level that can be attained.

159

Characteristic profiles developed by plotting the various parameters as a function of depth describe the site hydrologically. Characteristic profiles vary from year to year and between seasons within years. For this reason it is best to collect core samples at a season corresponding to minimum evapo-transpiration and maximum rainfall (determined from published climatological data). In the Southeast and Gulf Coast States the seasonal high groundwater table usually occurs in late December or early January.

This paper was written to provide a technique whereby one may define the nature of subsurface moisture characteristics. Particularly, the differentiation between a continuous groundwater table or aquifer and a perched or seansonal water table. It also defines the depth interval of an unsaturated zone with potential for receiving NOW materials. Obviously too thin a zone limits the utility of burial making other techniques more practical from an environmental and economic viewpoint.

The technique of describing subsurface hydrological characteristics from soil core data is advanced as an alternative to expensive geohydrological investigations utilizing groundwater monitoring wells. These investigations do not adequately describe perched water table conditions or soil vadose.

III-2   PROCEDURE

III-2.1   Soil Corings

Continuous soil cores are collected at the site to receive NOW materials. Several cores are taken to describe spatial variability. Cores may be taken seasonally to establish temporal variability. This would be recommended where climatological data is scant or non-existent.

III-2.2   Moisture Equivalents

The 1/3 and 15 bar moisture equivalents are determined on core samples segmented into 1 ft increments. Core samples are then split in half. The existing water content is determined by gravimetric weight loss which is the equivalent weight lost upon drying to constant weight at 105 C. The other half of the core sample is dried, ground and sieved to pass a 2 mm screen. Samples are saturated in contact with a ceramic plate and placed in a cell pressured at 1/3 bar and another at 15 bars. Soils are allowed to equilibrate at the pressures indicated followed by gravimetric moisture analysis of the sample by drying to constant weight at 105 C.

III-2.3   Porosity

Porosity is calculated as percent pore space by means of the following equation:

$$\text{Porosity, \%} = 100 - [\text{bulk density /particle density}] * 100$$

III-2.4   Bulk Density

The bulk density is measured as the ratio of mass to bulk volume. A paraffin clod technique gives an accurate assessment of strongly structured soils having aggregates of suitable size. Clods are often broken out from the core half used to determine existing moisture content. Loose or poorly structured soils are ground and sieved, then placed in a graduated cylinder to about 1/2 full. The soil is settled by gently tapping against the table top and volume occupied recorded. Contents are then weighed with bulk density reported as the mass per unit volume.

III-2.5   Particle Density

Particle density is a measure of the collective density of the solid particles. Dried and sieved materials are weighed into a volumetric flask. The flask is brought to approximately 1/2 its volume with water, followed by gentle boiling and agitation to remove entrapped air. Contents are cooled and brought to volume. The temperature of the water is recorded and flask weighed. Contents are removed from the flask and brought to volume with water and weighed again. Particle density is calculated as follows:

$$Ds = \frac{Dw(Ws - Wa)}{(Ws - Wa) - (Wsw - Ww)}$$

where

Dw = Density of water at observed temperature
Ws   = Weight of flask plus soil
Wa   = Weight of flask filled with air
Wsw = Weight of flask filled with soil and water
Ww   = Weight of flask filled with water at observed temperature

Results are generally presented by plotting the 4 parameters on a single graph, depth on one axis, moisture and porosity on the other. The area on the graph between the 15 bar moisture and the porosity is the range of moisture levels that could occur.

III-3   INTERPRETATION

At the point where percent moisture (as received moisture of the core sample) exceeds the 1/3 bar moisture equivalent a condition of free water exists. If the percent moisture exceeds 1/3 bar at a given depth but recedes to a value at less than the 1/3 bar equivalent deeper in the profile, then a perched water table is demonstrated (Figure A-III-1).

In the case where the percent moisture exceeds 1/3 bar and remains in excess, a continuous ground water table is demonstrated (Figure A-III-2). Depending upon the site and climatological conditions one may identify several discontinuous zones of saturation. It should be noted that the same physical constraints holding water in the profile serve to retard NOW constituent migration.

Burying NOW materials in a saturated zone in continuous contact with groundwater is a very undesirable condition and should be avoided unless constituents levels meet residential/farming upland soil criteria or the NOW materials are solidified to inert status.

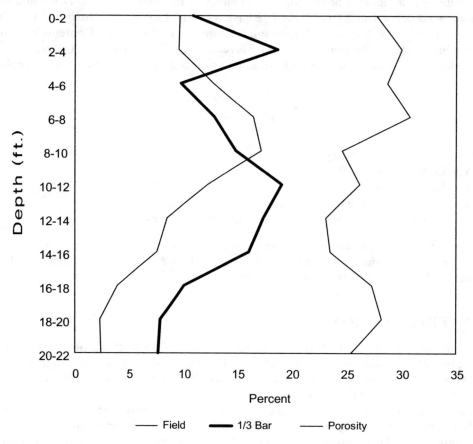

Soil Core Moisture Values and Porosity

Field   1/3 Bar   Porosity

A-III-1
MOISTURE PROFILE DEMONSTRATING A PERCHED WATER TABLE

Soil Core Moisture Values and Porosity

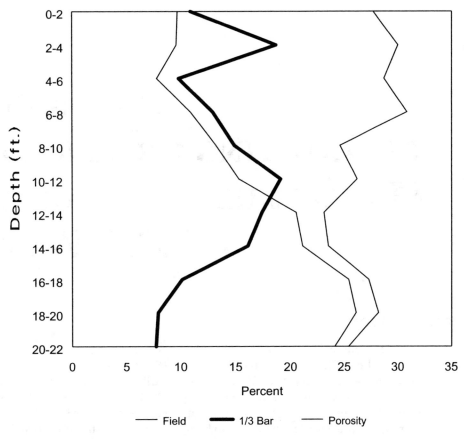

A-III-2
MOISTURE PROFILE DEMONSTRATING A GROUND WATER TABLE

APPENDIX IV
pH LIMITING CRITERIA FOR MANAGEMENT OF ONSHORE E&P WASTES

IV-1   INTRODUCTION

This appendix provides technical justification and guidance for pH limiting criteria for landspreading, burial, and roadspreading of Exploration and Production [E&P] wastes.

Resource Conservation and Recovery Act [RCRA] exempts drilling fluids, produced water and associated wastes from being classed as hazardous wastes [40 CFR Part 261]. Nevertheless, prudent operation requires knowledge of waste characteristics that make wastes hazardous. Waste pH equal to or less than 2.0 or equal to or greater than 12.5 exhibit the characteristic of corrosivity [40 CFR §261.22(a)(1)]. Low pH wastes can react to release from waste toxic levels of heavy metals, which otherwise are insoluble. Also, low and high pH waste pose a threat of chemical burns to humans and animals. E&P operations rely on both caustics and acids. However, less than 0.2 percent of E&P wastes would be classed as corrosive.

The Clean Water Act permit program requires effluent discharges to comply with pH limits, typically within the range 6-9 [40 CFR §133.102(c)]. However, only a limited number of onshore E&P discharges exist. EPA Region VI prohibits all discharges to surface water except Beneficial Use and Stripper operations.

This appendix recommends a pH range of 6-9 for roadspreading and on-site burial of wastes and a pH range of 6-8 for landspreading of wastes.

IV-2   pH AND RELATED CHEMISTRY

pH is the measure of hydrogen ion concentration $[H^+]$ of a substance. The negative base 10 logarithm of the $[H^+]$ concentration in moles/liter defines the pH:

$$pH = -\log [H^+] \tag{A-IV-1}$$

This relationship simplifies working with small numbers associated with $[H^+]$ in the natural environment. The range of pH goes from 0 to 14 as shown below:

H$^+$ moles/liter	10^{-1}	10^{-7}	10^{-14}
pH	1	7	14
reaction	acid	neutral	caustic

Soil reaction expresses the degree of acidity or alkalinity of the soil and is expressed in pH units.

IV-2.1   Hydrogen [H$^+$] and Hydroxide [OH$^-$] Ion Concentration

Water dissociates into hydrogen and hydroxide ions only slightly. Accordingly, water is known as a weak electrolyte. The dissociation is represented by the following equation:

$$H_2O = H^+ + OH^- \qquad \text{(A-IV-2)}$$

Pure water at room temperature [32° C] produces an ionization constant at equilibrium [Keq] of 10^{-14}:

$$Keq = \frac{(H^+)(OH^-)}{H_2O} \qquad \text{(A-IV-3)}$$

Since the activity of water [H$_2$O] is defined as unity, the equation can be simplified to express the fixed relationship between [H$^+$] and [OH$^-$] in pure water:

$$Keq = [H^+][OH^{-1}] = 10^{-14} \qquad \text{(A-IV-4)}$$

with [H$^+$] and [OH^{-1}] each being 10^{-7} mole/liter. It is fixed in that the sum of ion pair is always 10^{-14}. An increase in one ion necessitates a decrease in the other ion. This results in pH shift. For example, sodium hydroxide is a strong base which almost completely dissociates into sodium [Na$^+$] ions and [OH^{-1}] ions in solution. At 0.1 moles of NaOH /liter, [OH^{-1}] = 10^{-1}. Since [H$^+$][OH^{-1}] = 10^{-14} [Eq. A-IV-4], [H$^+$] equals 10^{-13}. The sodium has no significant effect on either [H$^+$] or [OH$^-$] in solution. The pH of the 0.1 moles/liter NaOH solution is:

$$\begin{aligned} pH &= -\log[10^{-13}] \\ &= -(-13) \\ &= 13 \end{aligned} \qquad \text{(A-IV-5)}$$

IV-3   HYDROLYSIS REACTORS AND BUFFERS

We define hydrolysis as the reaction of an ion with water to form [H$^+$] or [OH$^-$]. Hydrolysis increases either [H$^+$] or [OH$^-$] concentration above the normal 10^{-7} observed for pure water. Cations causing [OH$^-$] to decrease and [H$^+$] to increase is defined as acid hydrolysis. Anions causing [OH$^-$] to increase and [H$^+$] to decrease is said to undergo base hydrolysis: Aluminum [Al^{3+}] takes part in acid hydrolysis:

$$Al^{3+} + H_2O \rightarrow Al(OH)^{2+} + H^+ \qquad \text{(A-IV-6)}$$

since [H$^+$] increases.

For calcium [Ca^{2+}] the reaction increases [OH$^-$]:

$$CaCO_3 + H_2O \rightarrow Ca^{2+} + HCO_3^- + OH^-$$

and results in a base hydrolysis.

Anions of strong acids, e.g., [Cl$^-$] in hydrochloric acid, and cations of strong bases, e.g., Na$^+$ in sodium hydroxide do not hydrolyze in aqueous solutions. They form neutral salts [pH = 7] when combined, e.g., NaCl.

IV-4   ION EXCHANGE

Soil consists of positively and negatively charged particles. Cations in soil solution [hydrogen, H^+ aluminum, Al^{3+} calcium, Ca^{2+}] migrate and adsorb to negatively charged soil particles. Soils have a finite capacity to adsorb cations. This capacity is a function of soil type and is reversible, i.e., one cation can be exchanged for another. Thus, we refer this adsorption exchange ability of soils as Cation Exchange Capacity [CEC]. The amount and source of negatively charged particles depends on amount of organic matter present, clay content and soil reaction.

Since cations hydrolyze, the amount and speciation of cations [distribution] effects the ability of the soil to consume acid or bases, i.e., buffer capacity.

IV-5   SOIL BUFFERS

Materials that consume acids or bases while maintaining essentially constant pH are buffers. In soil, buffering can be provided by chemical equilibrium between $[H^+]$ in soil solution [active acidity] or $[H^+]$ or hydroxide cations [Al^{3+} and Fe^{3+}] adsorbed to soil particles [exchangeable acidity] and polymers of iron and aluminum hydroxyl compounds [non-exchangeable acidity]. A relatively constant pH results from the release or uptake of these components while maintaining the cation exchange equilibrium.

Cation exchange and hydrolysis provide the basis for buffering in soils. Changes in pH effect microorganisms and some plants. These pH changes result directly from changes in $[H^+]$ concentration or indirectly from chemicals, or nutrient deficiency or imbalances. Buffering capacity of soil pH effectively mitigates these problems.

IV-6   ANALYTICAL METHODS

pH is measurable for soils and waters. However, except for saturated soils, direct measurement of soil pH at equilibrium is not possible.

IV-7   WATER ANALYSIS FOR pH

Standard Method [4500-H^+ B.] commonly used for measuring pH of aqueous solutions relies on a direct reading instrument equipped with a combination glass indicating and calomel reference electrode. EPA guidelines [40 CFR §136, 49 Fed. Reg. 43260] require immediate pH analysis. While immediate analysis is desirable, laboratory determination is more accurate. Assuming the laboratory analysis is performed the same day as the sample is taken, laboratory determinations of pH are preferable.

The direct measurement of solid phase [solids] pH at equilibrium is not possible except under saturated conditions. Saturated soils under field conditions seldom occur, except in wetlands. Accordingly, moisture must be added for analysis. Adding water to the soil produces pH results not representative of interstitial pore liquid in soils at equilibrium.

Adding water to soil results in a metastable condition which changes with time. Accordingly, soil scientists devised a soil pH measurement at thermodynamic equilibrium. The factors affecting thermodynamic equilibria include:

- Spatial or materials variability
- Moisture content or soil/water ratio at which the pH readings are made.

Typically, the greater the soil/water ratio, the higher the pH reading [Jackson, 1967]. However, drying and grinding soil samples affect reaction kinetics but not the equilibria. Therefore, standardized sample preparation results in reproducible pH measurements, Annex A.

IV-8   WASTE ANALYSIS FOR pH

Mineral fraction, chemical additives and few soluble salts buffer E&P wastes, i.e., drilling mud. This buffering provides pH stability to the waste. Nevertheless, same day pH analysis assures accurate measurement.

We recommend collecting multiple samples [10 samples] from different areas of the waste area [pit]. Composite the samples into one *thoroughly mixed* single sample. This procedure controls sample variability. To assure consistency of measurements, dry [105° C.] and lightly grind to pass a 2 mm mesh sieve. Grinding the soil to a fine consistency increases mineral dissolution and shifts equilibria. This results in inaccurate and inconsistent pH measurements.

Moisten a representative sample of ground soil with distilled water to form a saturated paste. Refer to Appendix I, Annex H for Saturated Paste Extract procedure. Measure pH on the saturated paste extract, Appendix IV, Annex B.

IV-9   SOIL BUFFERS

Three conditions describe soil reaction:

- Acidic - pH <7
- Neutral - pH = 7
- Alkaline - pH >7

Precipitation percolation strips basic cations from soil. This process removes basic cations that buffer highly reactive H^+. Alkaline reaction occur in areas of low rainfall and high calcium, magnesium, sodium and potassium content [basic soils]. Few soils are neutral [pH = 7], the nearly ideal condition for plant-soil interaction.

IV-10   ACIDIC BUFFERS

Mitra, et al. [1963], Schwertmann and Jackson [1963], Coleman and Thomas [1964], Sawhney and Frink [1966] and Coleman and Thomas [1967] suggest existence of 3 buffer ranges in acid soil:

- Exchangeable H^+ attribute
- Exchangeable Al^{+3} attribute
- Hydrolysis Al^{+3} attribute

Exchangeable H^+ corresponds a pH <4.5 for soil suspensions in water. Exchangeable Al^{+3} occurs between pH 4.0 and 5.5. Hydrolysis of Al^{+3} occurs at pH >5.5 and produces $Al(OH)^{2+}$, $Al(OH)^+$ and polymeric Al-species.

The first and second buffer ranges are dominated by exchangeable Al^{+3} because of extreme reactivity of H^+ and dissolution of soil clay [Thomas, 1960]. Non-exchangeable polymeric Al-species form relatively weak buffers and amphoteric in nature [both acidic and basic] and serve as a sink for H^+ and OH^-.

IV-11   ALKALINE BUFFERS

Alkaline soils predominate in sub-humid and arid regions where insufficient rainfall develops to leach the soil profile [Mehlich, 1943]. Soils have pH values from about 6.5 to 7.4 are typically 100 percent base saturated and are considered neutral. These soils do not react to contact with strong acids. Weakly-to-moderately alkaline soils [7. ≤ pH ≤ 8.3] contain free calcium carbonate and react to contact with strong acids.

$$CaCO_3 + HCl \rightarrow CaCl_2 + H_2O + CO_2$$
(A-IV-8)

Strongly alkaline reaction, i.e., soils having pH > 8.5, result either from a critical percentage of exchangeable sodium [ESP > 15 percent] or free $Na_2 CO_3$ present [US Salinity Lab. Staff, 1954]. Dispersion and dissolution of soil humus often accompany soil reactions > pH 9.0 causing the surface to darken, giving rise to the term "black alkali soil" [Fireman and Wadleigh, 1951]. Soils remain in a relatively narrow pH range compared with many natural and anthroprogenic products, Figure A-IV-1. Also, soil pH conducive to optimum plant growth generally ranges from 5 to 8.5, with considerable variability noted in both endpoint and width of range between cultivars, Figures A-IV-2, 3 and 4.

IV-12   pH IMPACTS ON PLANTS

IV-12.1   Acidity Impacts

Extensive research has been conducted on the effects of low pH on plant-soil interaction [Hewitt, 1952; Olsen, 1958; Sutton and Halsworth, 1958; Rorrison, 1965; Clarkson, 1966; Jackson, 1967]. Direct impacts include:

- Injury by H^+
- Physiologically impaired absorption of plant nutrients
- Increased solubility of toxic ions.

Injury by H^+, i.e., first buffer range, is short term if the H^+ source is removed or corrected. In the long term and of greater significance is metal toxicity caused by increased solubility as a result of lowered pH and increased acid strength. Metals of concern include aluminum, manganese and iron.

Sparling [1967] developed solubility curves suggesting ferric iron causes no toxicity problem above pH 3.5. Similarly, he found aluminum and manganese typically cause no toxicity problem above pH 5.0. Soil having low oxygen levels i.e., low redox potential, make iron and manganese available in soil. Accordingly, the soil becomes toxic [Gotoh and Patrick, 1972]. Ferrous iron [Fe^{+2}] becomes soluble at about pH 6.0 under reducing conditions [redox potential E_h <+100 mV]. Toxic levels of iron [about 55 mg./l] result in marshland and waterlogged soils. Calcium and magnesium introduce antagonistic mechanisms that make the toxicity threshold for manganese less defined.

Increased solubility and decreased surface adsorption results from increased acidity, Figure A-IV-5. This elevates metal availability in soil solution.

IV-12.2   Fertility Impacts

Low pH directly reduces nutrient available in soil. Rorrison [1965] and Clarkson [1966] observed the reduced nutrients resulted from precipitation of phosphorus by aluminum in the soil. Similar precipitation occurs within the plant thereby reducing normal phosphorus metabolism.

Indirect impacts of low pH include impaired nitrification [microbiological conversion of ammonium to nitrate] and attach on plants by soil pathogens [Jackson, 1967]. Nitrate-nitrogen conversion in soil solution is efficient and rapid compared with ammonium-nitrogen conversion. In fact ammonium-nitrate conversion require roots release hydrogen ions that increases soil acidity.

IV-12.3   Alkalinity Impacts

Soil alkalinity can be as devastating to plants as acidity. Additionally, alkalinity is more difficult to correct. High pH reduces available micronutrients [iron, zinc, manganese and copper] to the point plant is undernourished [Hodgson, 1963; Adams, 1965; Grove and Ellis, 1980; Baker and Amacker, 1982]. Also, high pH results in loss of nitrogen since under high pH conditions ammonium converts to ammonia gas in the soil [Harsen and Kolenbrander, 1965].

Soil pH Conditions and Plant Community	pH Scale	Comparative Products
	0-	
	1-	Hydrochloric acid
		Phosphoric acid
	2-	Lemons
		Vinegar
	3-	Grapefruit
Barren		Apples
		Superphosphate
Extremely Acid Soils	4-	Tomatoes
Suitable for blueberries,		Acid rain
azaleas and rhododendrons		
	5-	Boric acid
Moderately Acid Soils		Fresh beans
Suitable for potatoes,		Distilled water
pine forests and grasses	6-	
Most row crops, vegetables and forage		Fresh corn
crops		Cows milk
Neutral soils	7-	
Alfalfa		Human blood
Slightly alkaline soils		
Hazard of deficiencies of Fe, Mn, Zn and	8-	Sea water
Cu		Calcium carbonate
Strongly alkaline soils		
Adapted species	9-	Bicarbonate of soda
Very strongly alkali soils		
Barren		
	10-	Milk of magnesia
	11-	Ammonia
		Washing soda
	12-	Trisodium phosphate
	13-	Sodium hydroxide
	14-	

FIGURE A-IV-1
SOIL pH CONDITIONS AND PLANT COMMUNITIES RELATIVE TO THE pH SCALE AND IN COMPARISON TO FAMILIAR PRODUCTS. (ADAPTED FROM: W. A. WAY, 1968)

Fiber, Grain and Special Crops

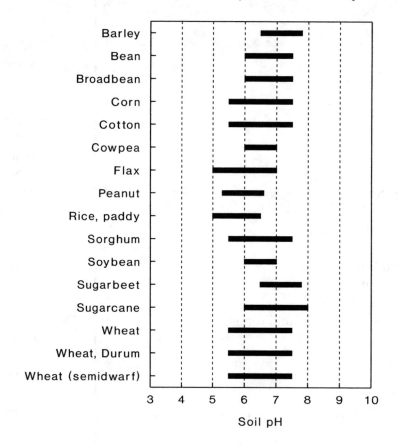

FIGURE A-IV-2
SOIL pH RANGE FOR OPTIMUM PLANT GROWTH

Grasses and Forage Crops

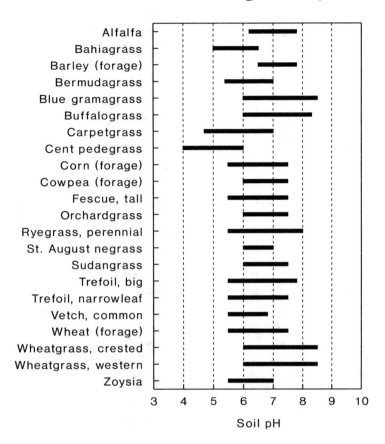

FIGURE A-IV-3
SOIL pH RANGE FOR OPTIMUM PLANT GROWTH

Vegetable and Fruit Crops

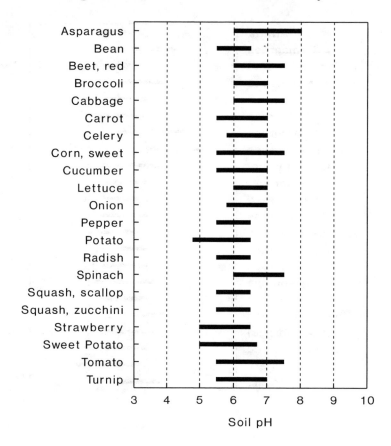

Soil pH

FIGURE A-IV-4
SOIL pH RANGE FOR OPTIMUM PLANT GROWTH

Metals Solubility as Influenced by pH

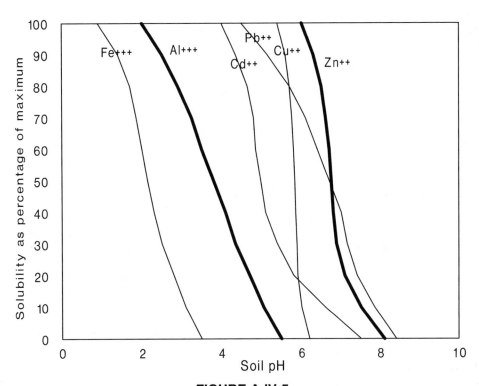

FIGURE A-IV-5
AFTER (SPARLING, 1972; LEGGETT, 1978; CAVARALLO AND MCBRIDE, 1980)

IV-13   SOIL pH AMENDMENTS

Soil amendments provide means of increasing or decreasing pH.

IV-13.1   Decreasing Soil pH

Acidifying amendments neutralize excess alkalinity and lower pH. Aluminum sulfate [$Al_2 (SO_4)_3$], ferrous sulfate [$Fe (SO_4)$] and elemental sulfur [S] result in lowered pH. Sulfur works well as a pH reducer. However, because it requires oxidation by soil microorganisms, it works more slowly than the other amendments. Waste acid works well and quickly. However, because of its reactivity, application to effect a desired loading rate into a soil volume of any size in somewhat uniform distribution requires specialized technique and careful management. Additionally, if the acid is truly a "waste" as defined by Resource Conservation and Recovery Act [RCRA], legal difficulties can result from discharging acid to the soil. The amount of acidifying agents is determined empirically by titration to a prescribed endpoint. Soil reaction with acids. Accordingly, rapid determination of the endpoint is essential to avoid overestimating the requirements.

IV-13.2   Increasing Soil pH

Lime [$CaCo_3$], the amendment of choice, reduces excess soil acidity. Lime serves two useful functions: (1) neutralize acidity [increases pH] and (2) adds calcium to the soil matrix.

Neutralization of excess acidity occurs at about pH 5.5. Addition of lime in excess of the exchangeable acid becomes necessary since the exchangeable acidity represents that portion of the total acidity replaced by neutral unbuffered salts.

Lime requirements [LR] based on near neutral pH results from adding a contribution for other forms of acidity. Based on resulting pH 6.8, equation A-IV-9 provides an estimate of the lime requirement, LR.:

$$LR, meq/100g = 2(\text{exchange acidity}) + 3[CEC/(pH)^2]$$

$$1 \ meq/100g = 1000 \ lb. \ CaCO_3/acre\text{-}6 \ in. \hspace{3em} \text{(A-IV-9)}$$

We multiply the exchange acidity by Equation A-IV-9 by 2 to compensate for the poor reaction of $CaCO_3$ in soil.

IV-14   pH CHARACTERISTICS OF E&P WASTES

E&P wastes consist of spent drilling fluids, produced waters and miscellaneous solids. The Resource Conservation and Recovery Act [RCRA] classifies these materials as non-hazardous waste under Subtitle C. EPA and API sampled and analyzed these wastes [EPA, 1987a; EPA, 1987b; API 1987] to determine if the exemption should continue.

IV-14.1   pH Test Results

EPA and API pH analyses of produced water samples showed these liquids to be nearly neutral on average Table A-IV-1. Reserve pit liquids were moderately alkaline. On the other hand, reserve pit solids show elevated pH levels, Table A-IV-1.

One produced water sample measured strongly acidic [EPA pH 3.0 and API pH 3.4]. Measured concentrations of iron [2200 mg/l] and manganese [120 mg/l] strongly suggest the low pH results from oxidation of reduced metals during anaerobic storage. The maximum pH measured by EPA and API [8.1 and 8.7, respectively] reflects bicarbonate/carbonate buffering, even though the produced water is sodium chloride, a neutral salt providing a neutral pH water.

Reserve pit liquids ranged from 6.5 [6.4] to 12.7 [13.0] for EPA [API] analyses, Table A-IV-1. The slightly elevated averages pH reflects the standard practice of treating drilling fluids with sodium hydroxide [NaOH] to retain solids in suspension.

Reserve pit solids ranged from 5.9 [7.9] to 12.8 [12.3] for EPA [API] samples. While the mean was 9.5 [10.2], Table A-IV-1. The differences between EPA and API pH values result from differences in sampling methods. EPA samples reflect surface conditions where greater potential for fixation of carbon dioxide [Holliday and Deuel, 1990]. This naturally acidifying reaction converts hydroxide to a moderately alkaline carbonate buffer system.

IV-15   ENVIRONMENTAL IMPACT OF pH

IV-15.1   Produced Water

Little or no short term effects occur when neutral pH produced water contacts soils having pH >5.5. This point emphasizes the importance of immediate soil remediation of produced water releases. Over time, soil pH increases proportionally with the amount of sodium adsorbed onto soil particles. Neutral pH brines contacting acid soils [pH <5.5] typically lowers soil pH 0.5 to 1.2 units more than the reaction from deionized water. As discussed previously, this reaction results from the exchangeable acidity present in the soil and the quantity of salt added by the release. In the long term, displacement of aluminum by sodium raises the pH markedly. Also, plant available phosphorus diminishes as a result of being precipitated by iron and aluminum in the soil.

Near-neutral pH produced water typically result in no adverse impact on pH in alkaline soils over the short term. Weakly calcareous soils can become sodic [ESP ≥ 15 percent] over the long term, with an anticipated increase in pH of about 2 pH units, as we shift from a $CaCO_3$ to a $(Na)_2CO_3$ buffered soil. Such a pH increase adversely impacts plants.

Acid water of low ionic strength have an impact similar to acid rain, i.e., decreases pH and increases leaching of nutrient cations in low buffer capacity soils [Galloway, et al., 1978]. On the other hand, neutral salt containing acid waters result in little to no adverse short term impact on acid soils. Injury to soils over time is measured by the quantity of sodium uptake. Interestingly, acid brines can have a beneficial impact on plants in calcareous soils when sodium content and ESP are controlled.

TABLE A-IV-1
pH SUMMARY STATISTICS FOR E&P WASTES SAMPLED IN PARALLEL BY EPA AND API

Test Statistic	Produced Waters		Pit Liquids		Pit Solids	
	EPA	API	EPA	API	EPA	API
Sample Size	17	17	15	15	19	19
Average	6.8	6.8	8.2	8.4	9.5	10.2
Median	7.2	6.9	7.6	7.9	9.0	10.3
Mode	7.7	6.8	7.3	7.6	8.8	9.0
Geometric Mean	6.7	6.7	8.1	8.2	9.4	10.2
Variance	1.4	1.2	3.3	3.6	2.8	1.4
Standard Deviation	1.2	1.1	1.8	1.9	1.7	1.2
Standard Error	0.3	0.3	0.5	0.5	0.4	0.3
Minimum	3.0	3.4	6.5	6.4	6.9	7.9
Maximum	8.1	8.7	12.7	13.0	12.8	12.3

IV-15.2   Reserve Pit Liquids

As demonstrated by Table A-IV-1, reserve pit liquids contain more alkalinity than produced water. This means pit liquids have a greater potential for adverse impact on weakly buffered soils than produced waters. Application of highly alkaline pit liquids to alkaline soils further hinders plant growth.

On the other hand, controlled applications of alkaline pit liquids to soils having pH <5.5 can be beneficial since excess acidity is neutralized. Also, this process increases the pH dependent cation exchange capacity and increases availability of plant nutrients.

Highly alkaline liquids [pH >12.5] offer a threat to human health and the environment because of the potential to cause caustic burns.

IV-15.3   Pit Solids

Reserve pit solids [water and oil base muds] analyses show 81.8 percent of the pH values fall between 5 and 8 units, Table A-IV-2. The presence of precipitated carbonates and hydroxides account for the strongly buffered solids. Most of the buffers are calcium and magnesium salts, which are not solubilized in the presence of sodium salts in the short term. Even in the long term, adverse impacts of reserve pit solids are low because of leaching of soluble sodium salts. This makes available calcium and magnesium to enhance soil fertility.

TABLE A-IV-2
FREQUENCY DISTRIBUTION FOR pH OF
WATER AND OIL BASE DRILLING MUDS

Cumulative Interval, pH	Frequency Distribution	Cumulative Percent
< 4.9	34	4.5
5.0 – 5.9	64	13.1
6.0 – 7.9	323	56.1
8.0 – 8.9	226	86.3
9.0 – 9.9	49	92.8
> 10	54	100.0
Total	750	

When applied to strongly acid soil, alkaline drilling fluids solids have a beneficial impact on plant growth because of increased soil pH [Miller and Pesaran, [1980]. However, excess amounts of alkaline solids can raise the pH above the acceptable range for plants, i.e., pH >9.

Soil pH values resulting from application of pit liquids and/or pit solids must match the land use, Table 7.2 or Table A-I-1. Excess acidity or alkalinity resulting from land disposal of E&P wastes can be corrected by adding lime or elemental sulfur, respectively.

IV-15.4   pH Criteria

pH 6-9 is satisfactory for wetlands areas but is too high for uplands areas, Table 7.2.

IV-15.5   pH Management

pH criteria depends on the disposal technique used. The techniques include, [Table A-IV-3]:

- injection
- road spreading
- land spreading

The disposal of E&P waste liquids using annular injection is an option, [Louviere and Reddoch, 1993]. Also, injection of liquids is acceptable in Class II wells, 40 CFR Part 146. No pH criteria are imposed on these injections. However, pH adjustment becomes necessary for landspreading and road-spreading, if the pH criteria is exceeded, Tables 7.2 and A-IV-3.

TABLE A-IV-3
SUMMARY OF E&P WASTE, DISPOSAL TECHNIQUE,
AND APPLICABLE pH CRITERIA

E & P Waste	Disposal Technique	pH Criteria
Liquids	deep well injection	NA*
	surface discharge	6 – 9
	roadspreading	6 – 9
	landspreading	6 – 8
Solids	annular injection	NA*
	landspreading	6 – 8
	burial or landfill	6 – 9

NA* – not applicable

Common acidifying agents used for reducing pH include aluminum sulfate ($Al_2(SO_4)_3$) and ferrous sulfate ($FeSO_4$). The agent of choice for increasing pH is calcium carbonate ($CaCO_3$).

Application rate of pH modifying amendments should be based on analytical results. Typically, landspread liquids are treated before application. Follow sampling and compositing recommendation included in Chapter 5. Unfortunately, poorly buffered soils can be over-treated, so it is important to take and analyze samples. Pit solids disposal can be accomplished by [Table A-IV-3].

- annular injection
- landspreading
- burial or landfill

Annular injection of drill solids is an accepted procedure [Louviere and Reddoch, 1993].

Landspreading provides opportunity for oxidation as the result of disking the solids into the soil. Accordingly, sulfur (S), which requires oxidation, is the amendment of choice.

Calcium hydroxide ($Ca(OH)_2$) or sodium hydroxide ($NaOH$) neutralize excess acidity to be buried or landfilled. Treated acidic waste to be landspread with calcium carbonate ($CaCO_3$).

IV-15.6   pH Adjustment Procedures

IV-15.6.1   Liquids.   pH adjustment of liquids is conducted in a tank prior to landspreading or disposal. Several examples of determination of the amount of amendments required are shown below.

IV-15.6.1.1   Example 1 - Reserve Pit Solids

An operator decides to apply 6,000 barrels of pit liquid to lease roads for dust control. The laboratory reported the pH of the liquids as 11 units. Rapid titration of the composite pit liquids yielded pH 7.0 at alkalinity of 15 meq (OH^-)/l. No other parameter such as salt or oil limits the application of the liquids to the roads. Calculate the quantity of aluminum sulfate ($Al_2(SO_4)_3$) required for alkalinity neutralization as follows:

$$15 \text{ meq}(OH^-)/l = 15 \text{ meq } H^+/l$$
$$Al_2(SO_4)_3 = [(27) * 2 + \{32 + 16 * 4\} * 3] = 342 \text{ g/mole}$$
$$Al \text{ has a valence} = 3$$
$$\text{g/equivalent} = 342 \text{ g/mole}/3 \text{ eq } H^+/\text{mole}$$
$$Al_2(SO_4)_3 = 114 \text{ g/equivalent}$$

$$\text{meq } H^+/\text{pit} = (15 \text{ meq } H^+/l) * 3.8 \text{ l/gal} * 42 \text{ gal/bbl} * 6000 \text{ bbl/pit}$$
$$= 14{,}364{,}000 \text{ meq } H^+/\text{pit}$$
$$\text{since 1 eq } H^+ = 1000 \text{ meq } H^+$$
$$\text{eq } H^+/\text{pit} = 14{,}364{,}000 \text{ meq/pit}/(1000 \text{ meq } H^+/\text{eq } H^+)$$
$$= 14{,}364 \text{ eq } H^+/\text{pit}$$
$$\text{amount } Al_2(SO_4)^3/\text{pit} = (14{,}364 \text{ eq } H^+/\text{pit}) * (114 \text{ g/eq})/454 \text{ g/lb}$$
$$= 3606 \text{ lb/pit}$$

IV-15.6.1.2   Example 2 - Produced Water

An acid frac results in 7000 barrels of fluid produced into frac tank. The operator decides to apply the liquid to lease roads for dust control. The liquid pH is 3.5 units. Rapid titration of a liquid sample results in 19 meq H^+/liter at pH 7.7. Calculate the amount of calcium carbonate [$CaCO_3$] required to neutralize the acidity.

$$19 \text{ meq }(H^+)/\text{liter} = 19 \text{ meq } OH^-/\text{liter}$$
$$Ca(CO_3) = [40 + \{12 + (16 * 3)\}] = 100 \text{ g/mole}$$
$$Ca \text{ valence} = 2$$
$$\text{g/equivalent} = (100 \text{ g/mole})/(2 \text{ eq } H^+/\text{mole})$$
$$Ca(CO_3) = 50 \text{ g/eq.}$$

$$\text{meq OH}^-/\text{tank} = (19 \text{ meq/l}) * 3.8 \text{ l/gal} * (42 \text{ gal/bbl}) * (7000 \text{ bbl/tank})$$
$$= 21,226,800 \text{ meq H}^+/\text{pit}$$

$$\text{Again 1 eq H}^+ = 1000 \text{ meq H}^+$$
$$\text{eq H}^+ \text{ pit} = 21,226,800/1000$$
$$= 21,227 \text{ eq X OH}^-/\text{tank}$$
$$\text{lb. CaCO}_3/\text{tank} = (21,227 \text{ eq OH}^-/\text{tank} * (50 \text{ g/eq})/(454 \text{ g/lb.})$$
$$= 2338 \text{ lb. CaCO}_3/\text{tank}$$

IV-15.6.2   Solids.   Buried or landfilled solids are pH adjusted in the pit prior to disposal. On the other hand, landspreadable solids are pH adjusted after application to the receiving soil. Neutralization reactions control pH adjustments. Neither acidity nor alkalinity limit the amount of solids. Assume pit solids require 4 acres to manage salt and hydrocarbons. A saturated paste extract pH for the solids equals 11.7 and alkalinity equals 6.7 meq OH$^-$/100 grams on a dry weight basis. The pit contains 11,500 barrels solids having a moisture content of 255 percent [dry weight basis] and a bulk density equaling 1.55 g/cm^3. The elemental sulfur required to reduce the pH of the solids is calculated as follows:

$$\text{Sulfur (S)} = (32 \text{ g S/mole})/(2 \text{ eq H}^+/\text{mole})$$
$$= 16 \text{ g S/eq H}^+$$
$$\text{wet solids, g} = [1.55 \text{ g/cc} * 1000 \text{ cc/l} * 3.8 \text{ l/gal} * 42 \text{ gal/bbl} * 11,500 \text{ bbl/pit}]$$
$$= 2.845 * 10^9 \text{ g/pit}$$
$$\text{dry solids, g} = (2.845 * 10^9)(100)$$
$$(100 + 255)$$
$$= 8.01 * 10^8 \text{ g/pit}$$

$$6.7 \text{ meq OH}^-/100 \text{ g} = 0.067 \text{ meq H}^+/\text{g}$$
$$\text{meq S/pit} = (0.067 \text{ meq S/g}) * (8.01 * 10^8 \text{ g/pit})$$
$$= (5.36^7 * 10^7 \text{ meq/pit}), \text{ but 1 eq} = 1000 \text{ meq}$$
$$\text{eq S/pit} = 5.367 * 10^7/1000 \text{ meq}$$
$$= 5.367 * 10^4 \text{ eq/pit}$$
$$\text{lb. S/pit} = (5.367 * 10^4 \text{ eq/pit}) * (16 \text{ g S/eq})/(454 \text{ g/lb.})$$
$$= 1891 \text{ lb/pit}$$
$$\text{lb. S/acre} = (1,891 \text{ lb/pit})/(4 \text{ acres/pit})$$
$$= 473 \text{ lb. S/acre}$$

REFERENCES

Adams, F. 1965. Manganese. In C. A. Black (ed.) Methods of soil analysis, Part 2. Chemical and microbiological properties. Agron. 9:1011–1018. Am. Soc. Agron., Inc. Madison, Wis.

API, American Petroleum Institute. 1987. Oil and Gas Industry Exploration and Production Wastes. DOC No. 471-01-09.

Baker, D. E. and M. C. Amacher. 1982. Nickel, copper, zinc, and cadmium. In Methods of soil analysis, Part 2. Chemical and microbiological properties. 2nd ed. (ed.) A. L. Page et. al. Agron. 9:323. Am. Soc. Agron., Inc. Madison, Wis.

Cavarallaro, N. and M. B. McBride. 1980. Activities of copper and cadmium in soil solutions as affected by pH. Soil Sci. Soc. Am. J. 44:729–732.

Clarkson, D. T. 1966. Aluminum tolerance in species within genus Agrostis. J. Ecol. 54:167–178.

Coleman, N. T. and G. W. Thomas. 1964. Buffer curves of acid clays as affected by the presence of ferric iron and aluminum. Soil Sci. Am. Proc. 28:187–190.

Coleman, N. T. and G. W. Thomas. 1967. The basic chemistry of soil acidity. In R.W. Pearsen and F. Adams (eds.): Soil acidity and liming. Agron. 12:1–41. Amer. Soc. Agron., Inc. Madison, Wis.

EPA, Environmental Protection Agency. 1987a. Report to Congress -Management of Wastes from Exploration, Development, and Production of Crude Oil, Natural Gas and Geothermal Energy.

EPA, Environmental Protection Agency. 1987b. Technical Report: Exploration, Development and Production of Crude Oil and Natural Gas: Field Sampling and Analytical Results. #530- SW-87-005, Appendix A-G.

Fireman, M. and C. H. Wadleigh. 1951. A statistical study of the relationship between pH and exchangeable sodium percentage of western soils. Soil Sci. 71:273–285.

Frink, C. R. and B. L. Sawhney. 1967. Neutralization of dilute aqueous aluminum salt solutions. Soil Sci. 103:144–148.

Galloway, J. N., E. B. Cowling, E. Gorham, and W. W. McFee. 1978. A national program for assessing the problem of atmospheric deposition (acid rain). A report to the Council on Environ. Qual. National Atmospheric Deposition Program NC-141. Washington, D.C.

Gotah, S. and W. H. Patrick, Jr. 1972. Transformation of manganese in a waterlogged soil as affected by redox potential and pH. Soil Sci. Soc. Am. Proc. 36:738–742.

Grove, J. H. and B. G. Ellis. 1980. Extractable iron and manganese as related to soil pH and applied chromium. Soil Sci. Soc. Am. J. 44:243–246.

Harmsen, G. W. and G. J. Kolenbrander. 1965. Soil inorganic nitrogen. In W. V. Bartholomew and F.F. Clark (ed.) Soil nitrogen. Agron. 10:43–92.

Helling, C. S., G. Chesters, and R. B. Corey. 1964. Contributions of organic matter and clay to soil cation exchange capacity as affected by the pH of the saturating solution. Soil Sci. Soc. Am. Proc. 28:517–520.

Hewitt, E. J. 1952. A biological approach to the problems of soil acidity. Trans. Int. Soil Sci. Jt. Meet. Dublin 1:107– 118.

Hodgson, J. F. 1963. Chemistry of micronutrient elements in soils. Advan. Agron. 15:119–159.

Holliday, G. H. and L. E. Deuel, Jr. 1990. A statistical Review of API and EPA Sampling and Analysis of Oil and Gas Field Wastes, Society of Pit Engineers, SPE 20711.

Jackson, W. A. 1967. Physiological effects of soil acidity. In Soil acidity and liming. R.W. Pearson and F. Adams (ed.)Agron. 12:43–124. Am. Soc. Agron., Inc. Madison, Wis.

Leggett, G. E. 1978. Interaction of monomeric silicic acid with copper and zinc and chemical changes of the precipitates with aging. Soil Sci. Soc. Am. J. 42:262–267.

Louviere, R. J. and R. J. Reddoch. 1993. Onsite Disposal of Rig Wastes via Slurrification and Annular Injection, SPE/AIDC 25755.

Mehlich, A. 1943. The significance of percentage base saturation and pH in relation to soil differences. Soil Sci. Soc. Amer. Proc. 7:167–174.

Miller, R. W. and P. Pesaran. 1980. Effects of drilling fluids on soil and plants: II. Complete drilling fluid mixtures. J. Environ. Qual. 9:552–556.

Mitra, R. P., B. K. Sharma, and B. S. Kapoor. 1963. Three stages in the titration of montmorillonite in water and in acetonitrile-benzene mixtures. Ind. J. Chem. 1:225–226.

Olsen, C. 1958. Iron uptake in different plant species as a function of the pH value of the nutrient solution. Physiol. Pl. II: 889–905.

Rorison, I. H. 1965. The effect of aluminum on the uptake and incorporation of phosphate by excised Sainfoin roots. New Phytol 64:23–27.

Sparling, J. H. 1967. The occurrence of Schoenus nigricans L. in blanket bogs. II. Experiments on the growth of S. nigricans under controlled conditions. J. Ecol. 55:15–31.

Sutton, C. D. and E. G. Hallsworth. 1958. Studies on the nutrition of forage legumes. I. Toxicity of low pH and high manganese supply to Lucerne, as affected by climatic factors and calcium supply. Pl. Soil 9:305–317.

Schwertmann, U. and M. L. Jackson. 1963. Hydrogen-aluminum clays: A third buffer range appearing in potentiometric titration. Sci. 139:1052–1053.

Soil Salinity Lab. Staff. 1954. Diagnosis and improvement of saline and alkali soils. USDA Handbook 60. U.S. Printing Office, Washington, D.C.

Standard Methods for the Examination of Water and Waste Water, 18[+L] Ed. (1992) Method 4500-H[+B]. Electrometric Method.

Way, W. A. 1968. The whys and hows of liming. U of Vt. Brieflet 997.

ANNEX A
E&P SAMPLE PREPARATION

1.0. Scope and Application
   1.1. This method is used to prepare samples for analysis by the protocols listed below:
      1.1.1. Saturated Paste Extract
      1.1.2. pH
      1.1.3. Active Acidity (Lime Requirement)
      1.1.4. Alkalinity (Acid Requirement)

2.0. Summary of Method
   2.1. The sample is homogenized, dried at 105°C and ground prior to the individual analyses.

3.0. Apparatus and Materials
   3.1. Oven capable to 105°C (+/− 2°C)
   3.2. Grinding apparatus
   3.3. Drying pans
   3.4. Balance

4.0. Procedure
   4.1. Homogenize the sample thoroughly.
   4.2. Weigh a pan to the nearest 0.1 g that is large enough to hold 250 g sample.
   4.3. Weigh 100 to 200 g homogenized sample to pan, and place pan in oven at 105°C until a constant weight is achieved. Record weights to calculate moisture content.
   4.4. Grind the material so that it will pass a 2-mm sieve. Sample is now ready for the analyses listed.

5.0. Procedure for Hydrophobic Material
   5.1. Tests for hydrophobicity
      5.1.1. Visible blobs of oil or grease
      5.1.2. The sample presses into a single damp looking mass when crushed with mortar and pestle and will not hydrate with water.
      5.1.3. Sample leaves an oily mark when pressed between two pieces of filter paper.
      5.1.4. Sample feels damp when pinched between fingers.
   5.2. Place sample in muffle furnace and heat to 250°C for 1 hr.
   5.3. Raise temperature to 350°C at 50°C intervals allowing smoke to dissipate between adjustments. Do not allow sample to catch fire or exceed 390°C.
   5.4. Cool the sample and grind it to pass 2-mm sieve. The sample is now ready for the appropriate analyses.

6.0. Calculation
   6.1. Moisture Content (dry weight basis)

$$\text{Moisture, \%} = (W - D)/(D - P) * 100$$

where

W = wet weight of sample + pan, g
D = dry weight of sample + pan, g
P = weight of pan, g

ANNEX B
pH DETERMINATION

1.0. Scope and Application
   1.1. This method is an electrometric procedure that provides a direct measurement of pH in water, soils and E&P wastes.

2.0. Summary of Method
2.1. The pH of E&P liquids and solids are measured in the field whenever possible.
2.2. The dried and prepared sample (soil or solid waste) is mixed with type II water.

3.0. Interferences
3.1. Samples with very low or very high pH may give incorrect readings on the meter.
3.2. Temperature fluctuation will cause measurement errors.
3.3. Errors will occur when the electrodes become coated with oily materials. The electrode can either (1) be cleaned with an ultrasonic bath, or (2) be washed with detergent, rinsed with water, followed by 1:10 HCl, then deionized water.

4.0. Apparatus and Materials
4.1. Temperature compensating pH meter.
4.2. Electrodes:
    4.2.1. Calomel reference electrode.
    4.2.2. Glass indicating electrode.
    4.2.3. Or a combination glass-calomel electrode.

4.3. 50 ml beaker or other suitable container.

5.0. Reagents
5.1. ASTM Type II water (ASTM Dll93).
5.2. Standard buffers, usually pH 2, 4, 7, 10 and 12.

6.0. Sample Preparation
6.1. See sample preparation section (ANNEX A).

7.0. Procedure
7.1. Calibration:
    7.1.1. Set up instrument according to the manufacturers instruction. Operational function and settings are to varied to be covered in this generalized method.
    7.1.2. Standardize instrument/electrode systems against buffers that bracket the expected pH of the sample(s).

7.2. pH measurement of samples:
    7.2.1. Insert electrodes into aqueous solutions or E&P liquids and read pH immediately on the standardized system.
    7.2.2. Mix or stir E&P waste suspensions to be measured in the field or as received for 10 sec, then insert electrodes and read pH immediately on the standardized system.
    7.2.3. Stir saturated paste for 10 sec, then insert electrodes and read pH immediately on the standardized system. A glass rod attached to the electrodes or tip cover will serve to protect electrodes, as will a slight swirling action, on insertion.

8.0. Quality Control
8.1. Duplicate samples and check standards should be analyzed routinely.
8.2. Rinse electrodes between samples.

9.0. References
9.1. EPA method 9045. Test Methods for Evaluating Solid Waste-Physical/Chemical Methods. EPA SW-846, 1986, revised.
9.2. Method 12-2.6. Glass Electrode-Calomel Electrode pH Meter Method. Methods of Soil Analysis. Part 2-Chemical and Microbiological Properties. A. L. Page (ed.).Agron. 9. Amer. Soc. Agron. Madison, Wis. 1982.

ANNEX C
ALKALINITY (ACID REQUIREMENT)

1.0. Scope and Application

    1.1. Alkalinity is a measure of the base forms in E&P waste liquids or solids that are readily neutralized on titration with a strong acid. The acidity requirement is a measure of the amount of acid needed to lower the waste matrix pH to < 8.0.

2.0. Summary of Method

    2.1. Alkalinity is measured by potentiometric titration with hydrochloric acid (HCl) used as the titrant and reported in meq OH^-/liter for liquids or meq OH^-/100 g for solids.

    2.2. E&P waste liquids are titrated direct with a pH meter and appropriate electrodes.

    2.3. E&P waste solids are titrated in-situ after preparing a dilute suspension with Type II water.

3.0. Interferences

    3.1. Exposure to air and fixation of CO_2. This is overcome by rapid titration to pH 6.5.

4.0. Apparatus

    4.1. Temperature compensating pH meter.

    4.2. Electrodes

        4.2.1. Calomel reference electrode.

        4.2.2. Glass electrode.

        4.2.3. Or a combination glass-calomel electrode.

5.0. Reagents

    5.1. ASTM Type II water (ASTM D 1193).

    5.2. pH 4 and pH 10 calibration buffer solutions.

    5.3. 0.1 N hydrochloric acid (HCl)

    5.4. 50 ml burette.

6.0. Procedure

    a. E&P Waste Liquids

        6.1. Titrate a suitable aliquot of E&P waste liquid such that a minimum 20 ml of the 50 ml burette is utilized to the potentiometric endpoint (pH 6.5).

        6.2. The pH is monitored on a direct reading instrument with sample under constant agitation while adding acid.

    b. E&P Waste Solids

        6.1. Weigh 10 g E&P waste solid sample into 250 ml wide mouth erlenmeyer flask or tall form beaker.

        6.2. Add 150 ml Type II water, and insert electrodes for immediate titration.

        6.3. Titrate under constant agitation to potentiometric endpoint (pH 6.5).

7.1. Calculations

    7.1. Alkalinity (Acid Requirement)

        a. E&P Liquid (meq OH^-/liter)

        meq OH^-/liter = A * B * 1000 / ml sample

        where:

                A = ml titrant used

                B = normality (N) of titrant

        b. E&P Solid (meq OH^-/100 g)

        meq OH^-/100 g = A * B * 100 / g sample

where:

A = ml titrant used

B = normality (*N*) of titrant

g sample (dry) = [g sample (wet) * 100]/[100 + % moisture]

ANNEX D
ACTIVE ACIDITY (LIME REQUIREMENT)

1.0. Scope and Application

    1.1. Active acidity is a measure of the acid forms in E&P waste liquids or solids that cause elevated hydrogen ion (H^+) activities and pH values < 6.0. The lime or base requirement is a measure of the amount of base needed to raise the pH to > 6.0.

2.0. Summary of Method

    2.1. Active acidity is measured by potentiometric titration using sodium hydroxide as the titrant and reported in meq H^+/liter for liquids or meq H^+/100 g for solids.

    2.2. E&P waste liquids are titrated direct with a pH meter and appropriate electrodes.

    2.3. E&P waste solids are titrated in-situ after preparing a dilute suspension with Type II water.

3.0. Interferences

    3.1. Aluminosilicate minerals associated with E&P waste solids can consume erroneously high quantities of base by neutralization of polymerized species and/or formation of aluminates when titrated too slowly. This is overcome by rapid titration and holding the pH 7.5 end point 30 sec.

4.0. Apparatus

    4.1. Temperature compensating pH meter.

    4.2. Electrodes

        4.2.1. Calomel reference electrode.

        4.2.2. Glass electrode.

        4.2.3. Or a combination glass-calomel electrode.

5.0. Reagents

    5.1. ASTM Type II water (ASTM D 1193).

    5.2. pH 4 and pH 10 calibration buffer solutions.

    5.3. 0.1 *N* sodium hydroxide (NaOH).

    5.4. 50 ml burette.

6.0. Procedure

    a. E&P Waste Liquids

        6.1. Titrate a suitable aliquot of E&P waste liquid such that a minimum 20 ml of the 50 ml burette is utilized to the potentiometric endpoint (pH 7.5).

        6.2. The pH is monitored on a direct reading instrument with sample under constant agitation while adding base.

    b. E&P Waste Solids

        6.1. Weigh 10 g E&P waste solid sample into 250 ml wide mouth erlenmeyer flask or tall form beaker.

        6.2. Add 150 ml Type II water, agitate and allow to stand 30 min.

        6.3. Insert electrode(s) and titrate under constant agitation to potentiometric endpoint (pH 7.5).

7.0. Calculations
   7.1. Active Acidity (Lime Requirement)
      a. E&P Liquid (meq H$^+$/liter)

        meq H+/liter = A * B * 1000 / ml sample
        where:
            A = ml titrant used
            B = normality (N) of titrant

      b. E&P Solid (meq H$^+$/100 g)

        meq H$^+$/100 g = A * B * 100 / g sample
        where:
            A = ml titrant used
            B = normality (N) of titrant
      g sample (dry) = [g sample (wet) * 100]/[100 + % moisture]

PLATES

PLATE 1
SILICA TETRAHEDRON

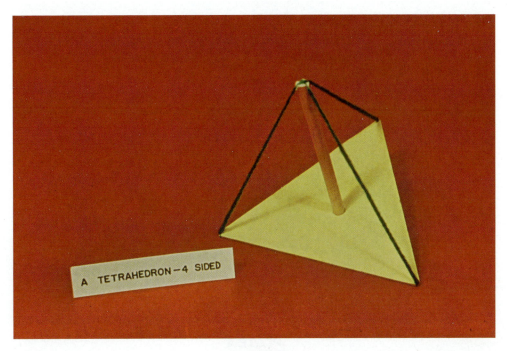

PLATE 2
TETRAHEDRON

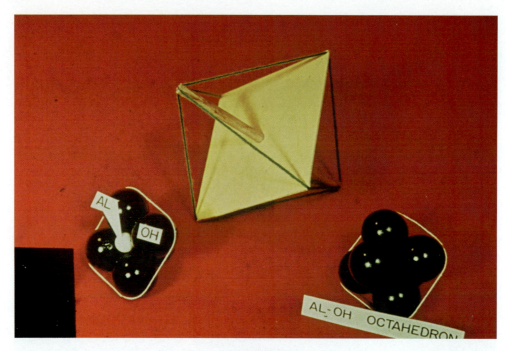

PLATE 3
ALUMINUM OCTAHEDRON

PLATE 4
OCTAHEDRAL ARRANGEMENT OF ALUMINUM HYDROXIDE IN KAOLINITE AND GIBBSITE

PLATE 5
CHEMICAL AND STRUCTURAL FEATURES OF KAOLINITE

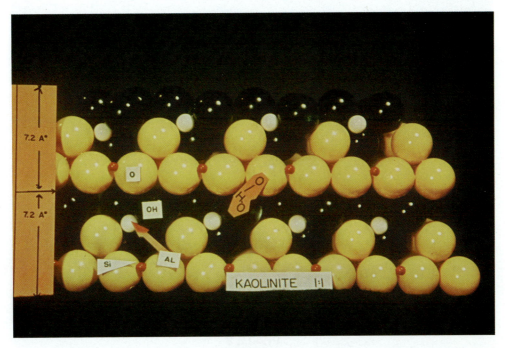

PLATE 6
HYDROGEN BONDING BETWEEN LAYERS OF KAOLINITE CLAY

PLATE 7
ILLITE CLAY

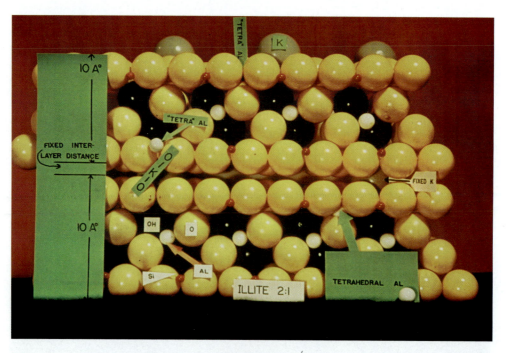

PLATE 8
SMALL FIXED INTERLAYER SPACE OF ILLITE PREVENTS WATER AND CATIONS
FROM ENTERING

PLATE 9
SUBSTITUTION OF MAGNESIUM FOR ALUMINUM IN MONTMORILLONITE INCREASES CEC

PLATE 10
THE LOW ATTRACTION BETWEEN OXYGEN ATOMS IN ADJACENT LAYERS OF MONTMORILLONITE ALLOW FREE MOVEMENT OF WATER AND EXCHANGE OF CATIONS

PLATE 11
CATIONS SUCH AS MAGNESIUM AND CALCIUM ARE HYDRATED BY
6 MOLECULES OF WATER

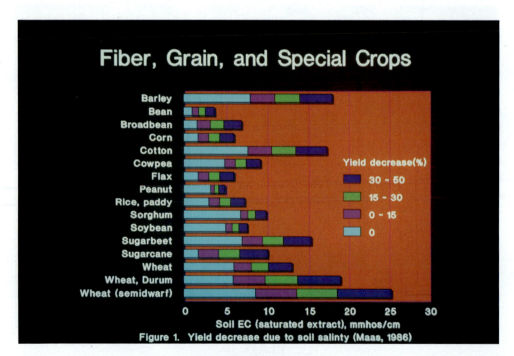

PLATE 12
YIELD DECREASE DUE TO SOIL SALINITY

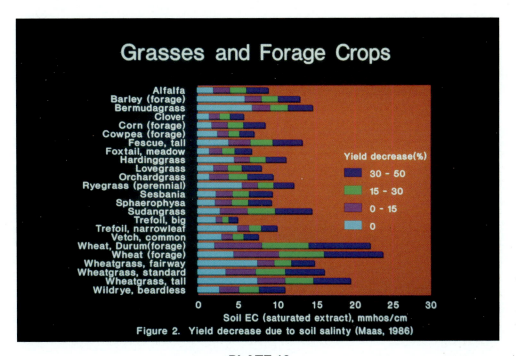

PLATE 13
YIELD DECREASE DUE TO SOIL SALINITY

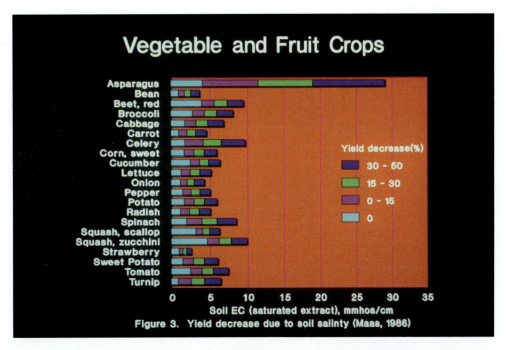

PLATE 14
YIELD DECREASE DUE TO SOIL SALINITY

PLATE 15
SAMPLING LARGE PRODUCTION FACILITY PIT IN INDONESIA

INDEX